GIVING THE FINGER

GIVING THE FINGER
Risking It All to Fish the World's Deadliest Sea

SCOTT CAMPBELL JR.

with Jim Ruland

LYONS PRESS
Guilford, Connecticut
An imprint of Globe Pequot Press

To buy books in quantity for corporate use
or incentives, call **(800) 962-0973**
or e-mail **premiums@GlobePequot.com.**

Lyons Press is an imprint of Globe Pequot Press.

All photos are by the author or are part of the author's collection.

Text Design: Sheryl P. Kober
Layout artist: Justin Marciano
Project editor: Ellen Urban

Library of Congress Cataloging-in-Publication Data is available on file.

ISBN 978-0-7627-9131-6

Printed in the United States of America

10 9 8 7 6 5 4 3 2 1

Dedicated to Keith Criner, aka Moose, and his family.

CONTENTS

CHAPTER ONE

DEAD IN THE WATER

"Honey, I had a little bit of an accident."

I was sitting in the galley of the *Amatuli* in the middle of the Bering Sea during a ferocious winter storm. Talking on the satellite phone with my wife, I looked down at the ring finger on my right hand.

Or what was left of my ring finger.

"What happened?" Lisa asked.

She probably figured that if I was calling her on the phone during crab season, it had to be serious. It *was* serious, but I didn't want her to know that.

"Oh, I hurt my finger."

"What did you do?"

"I banged it up a little bit." I explained how I'd smashed my finger down in the engine room, and that it wasn't too bad, just the tip of my finger really.

"You mean like around the nail?"

"Yeah," I said. "Like that."

But it was more than that. A lot more.

The nail was gone. So was the flesh around it.

The only thing left was the bone sticking out of my finger.

And blood. Lots and lots of blood.

The fingers of the human hand, I would soon learn, are made up of three distinct parts called phalanxes: the proximal, middle, and distal. The proximal phalanx is where the finger attaches to the hand. The distal is the tip. So, technically, I wasn't lying to my wife when I told her that I'd hurt the tip of my finger. But the distal phalanx was more than hurt. It was missing. All the way down to the knuckle. Just gone.

Lisa was a nurse. She knew all the medical mumbo jumbo. But if she knew what I knew, could see the mess that I was looking at, she'd make me come home, which I'd already decided I didn't want to do.

"Scott, what happened?"

Good question.

The strike happened. The ice happened. The shitty air compressor on the shitty *Amatuli* happened. All the things that can go wrong when you go crab fishing on the Bering Sea happened.

I'd been around crab boats all my life, but this was my first time running one professionally. I'd been fishing with my father and my uncle from the time I was fifteen years old and around boats in one way or another since I was four. The sea was in me, a part of who I was, but for the first time in my life, I was the captain. My crew. My boat. I'd finally stepped out of my father's shadow, and now it was *my* reputation that was on the line. I had a lot of pressure on me, and I was ready for the challenge—at least I thought I was. But my mangled hand told a different story.

The *Amatuli* was a crappy old junker, long and skinny with a house aft instead of forward and a tendency to roll. She wasn't designed for extreme weather. In fact she had no business being up there in the Bering Sea's blisteringly cold cauldron.

The *Amatuli* was the type of boat you got when you were first starting out. They don't give young guys new boats and a half a million dollars worth of gear and say, "Go catch some crab!" You have to earn it.

When the owner of the *Amatuli* asked me to skipper his boat in 2000, it was up to me to hire the crew. I called every fisherman I'd ever worked with and begged him to give me a chance. "You know what I can do on deck," I told each one of them. "I'll do the same thing in the wheelhouse. But I gotta have a good crew to make it happen."

Though skeptical at first, one by one they came around. Eventually I was able to convince enough guys to put together a halfway decent crew. My best crewman was Jerry Perkey, a longtime friend and super-skilled deckhand who had an offbeat sense of humor. A dependable guy. He even saved my life once.

I was able to fill all the positions on the boat except one. I couldn't find an engineer to help keep the boat running smoothly while I focused on fishing. I decided that I'd serve as my own. It wasn't an ideal situation, but I didn't have many options.

My strategy that season was to fish with two loads of gear. We'd leave one set of crab pots in the water as an underwater storage, come back to port, and get the second set on deck. Then, once the season started, we'd go back out to the fishing grounds, set the second set of pots, and start hauling the first set.

Double the work. Double the reward.

The crew assembled in Dutch Harbor, Alaska, where the *Amatuli* was waiting for us to make her seaworthy. But she was in much worse shape than I thought. We spent a ton of time doing basic maintenance on the boat, things that hadn't been done in who knows how long. When we went out to sea to put our first set of gear in the water, I discovered how unstable she was.

She was always breaking down. I'd set a string, and something would go wrong. Set another string, and something else would fail. I could hardly get anything done, because the boat was constantly falling apart on me. I've always been handy with engines, so I was able to patch it together, but without an engineer on board, I spent as much time in the engine room as I did in the wheelhouse.

When we got back to Dutch Harbor, I received word that negotiations with the canneries were underway. That meant that a strike was coming. Strikes were a fairly common occurrence back then. If the fishermen didn't like the price the canneries were offering, we'd stand down and try to get a better deal. Normally a strike lasted a few days. This one lasted three weeks.

3

Years after my stint on the *Amatuli*, still hauling crab pots and trying to get them aboard before the ice floes got them.

There I was, my first crack at being a captain, and I had to stand down for almost a month waiting for my opportunity.

And then the ice started coming down.

The sea ice forms in the northern part of the Bering Sea and gets blown southward. The ice is as much a part of the sea as the wind and the waves, but you can't fish in it, and you sure as hell can't leave your gear in it.

I had a quarter of a million dollars' worth of gear in the water that was going to get lost to the ice, which would make the owner of the *Amatuli* very unhappy. There was no choice but to take my second load of gear off the *Amatuli*, go back up to the fishing grounds, and try to get the gear out before the ice did.

Some people might say, "Relax. It's out of your control." But that's not me. I was mad at myself. I'd underestimated how long the strike would last and miscalculated how far the ice would come down. I'd made two costly errors in judgment, and we hadn't even started fishing yet.

How many more mistakes could I afford to make?

Answer: none.

———

We were about twenty-four hours out when the air compressor died.

The air compressor powers the pneumatic throttle. Without compressed air, the screws won't turn. If the propellers don't turn, you can't make way. If you're not making way, you're a sitting duck. As any hunter will tell you, a sitting duck is a dead duck.

The difference between riding the swells and getting pummeled by them is night and day. The action of the waves is more violent when your boat's not motoring. You pitch forward and back, roll side to side. Everything in the boat gets tossed around. Equipment that has been safely stowed and secured shakes free and crashes around on the deck. It's dangerous and disorienting. When you're sitting dead in the water, you're confronted with the sea's enormous power. It's a reminder that a few inches of fiberglass and steel are the only things that separate you from a watery grave.

It was the middle of the night. Fiercely cold. Thirty-foot seas. Shrieking winds blowing out of the north. We were going sideways into the swells. I had to get that compressor fixed—and fast.

The engine room was loud, hot, and reeking of diesel. I'd been spending so much time down in that hellhole that I practically had every inch of it burned into my memory. There in the back corner was the compressor, and I quickly identified the problem.

The bolts that secured the compressor's motor to the platform had vibrated loose. The motor was shifting forward and back with the motion of the waves. Every time it shifted forward, it put enough slack in the belt to stop the works. While I had my face in the compressor, trying to figure out the quickest way to bolt the sucker down, a huge wave slammed into the boat. I tried to brace myself, but there was nothing to hold on to. As the *Amatuli* rolled over onto its side, I fell right on top of the air compressor, my weight pulled the slack out of the belt, and the motor started up again while my hand was still in the works. My hand got thrashed around a little bit, but I pulled it out right away. It stung like hell.

I tucked my hand under my armpit. And while I was being tossed around like a boot in a dryer, I failed to grasp what exactly had happened.

I put my foot on the motor and got the bolts retightened to the point where I thought it would hold. After I got the air compressor working and power to the throttle was restored, I made my way topside.

I felt really hot. I could feel the sweat dripping off me.

God, I'm sweating like a pig, I thought.

I went up the ladder, and as I pulled myself up, I caught sight of my hand and couldn't believe what I was seeing.

It wasn't sweat I'd felt, but blood, and it was squirting all over the place.

I'm not particularly squeamish, but this was bad. A closer look at my finger revealed a mangled mess. Something had grabbed hold of the skin and pulled it off. The nail was gone. So was the flesh. I was looking right at the bone.

I wanted to run screaming, but where would I go? Who would I go screaming to?

I was the captain. There was no one I could go to for help. It was up to me to figure out what to do.

I wrapped my finger in a T-shirt and went up to the galley where the crew was assembled.

"I've got good news and bad news," I said. "The good news is I fixed the compressor."

They looked pretty relieved. Simple math will tell you that whenever you take a roll that's more than forty-five degrees, there's a chance you won't recover. Another roll like the one that sent me flying into the compressor and we could very well have ended up at the bottom of the Bering Sea. So they were all on edge, especially the new guys.

"But we've got a little bit of an issue."

I showed them my finger.

Everyone freaked. There were some experienced fishermen on the boat, but none of them had seen an exposed bone before.

"What the hell happened?" Jerry asked.

I told him what went down in the engine room, and then I laid out my plan. We were six hours from getting the gear. We were going to go up and finish our job and get the gear moved to safety. Then we were going to turn around and go back to Dutch Harbor as planned.

"You're delirious!" Jerry shouted. "You're in so much pain you don't know what you're saying!"

"Guys," I said, "there's nothing we can do. By the time we get back to port, we're not going to be able to save my finger."

"We gotta turn around," Jerry insisted.

"No, we're gonna keep going," I said as calmly as I could. "The damage is done. There's nothing we can do about my finger, but we can save the gear."

What I didn't tell them was that there was no way in hell I could go back to Dutch Harbor and face my father and my uncle with my tail between my legs. That would be the end of my career as a captain. It would be over before I caught my first crab.

"Where is it?" Jerry asked.

"Where's what?"

"Your finger."

"Gone."

They wanted to go look for it, to see if it could be reattached, but there was no point. It was probably floating in the bilges somewhere or flushed out to sea. Though I loved the excitement and the danger, I'd been cursing that sea my whole life for all the things it had taken from me. If I turned around and went home, I could add one more thing to the list. By staying and finishing the job, I was giving the finger to the Bering Sea, and it was damn well going to give me something back.

CHAPTER TWO

PROTECTED WATERS

THE FIRST TIME I WENT OUT TO SEA, I GOT SERIOUSLY SICK. IT WAS MY first time for a lot of things: my first trip on an airplane, my first visit to Alaska, and the first time I met my dad.

I had just turned four when my grandparents took me up to Kodiak, Alaska, to go live with my dad and his wife—my new mom. I knew my dad was a fisherman, but that's *all* I knew.

My dad grew up in Milton-Freewater in the Walla Walla Valley in northeastern Oregon. Milton-Freewater sits just south of the Washington border across from the city of Walla Walla, which is where I make my home today. It's an odd place for a fisherman to live, because it's nowhere near the ocean. It's closer to Idaho than it is to the Pacific.

The Walla Walla Valley has always been farm country, but today it's known for its vineyards and wineries. My dad's father was a farmer who raised cherries, apples—you name it. My grandpa was a Navy man who'd served in the South Pacific during World War II. He didn't see any combat, but he raised his three sons with great discipline. He was a strict, squared-away man who was always clean-shaven and lived his life according to rigorous principles, which he instilled in his three sons. Or tried to anyway. One thing you should know about us Campbells: We never do things the easy way.

For the first four years of my life, the Walla Walla Valley was my home, but I don't remember much about those days. My dad and my

From left to right: Uncle Kevin Campbell, holding his oldest son Alan; Uncle Dan Campbell, holding his son Chris; my grandfather, Howard Campbell; and my father and me.

biological mother weren't married and didn't have much of a relationship. My dad liked to work. My biological mother liked to party. They never had the same priorities, especially when it came to raising me.

My dad took off to Alaska before I was even born. After my biological mother gave birth to me, she called my dad's parents from the hospital to let them know that they were officially grandparents. They were pretty shocked. No one had told them that their son had a pregnant girlfriend.

At first my biological mother and I lived in Milton-Freewater with my grandparents. They probably hoped that she'd work things out with my dad, and that they'd "do the right thing." But there was no love between my dad and my biological mother, and we moved out.

We moved around the valley a lot. My biological mother raised me for the first few years, but she had serious issues with drugs and alcohol. Being a full-time single mother was more than she could handle. By the time I was three, I was back with my grandparents in Milton-Freewater, and that's where I stayed until shortly after my fourth birthday in May 1978.

Earlier that year my dad had gotten married and bought a home up in Kodiak, Alaska. He was finally settling down, and he wanted me to come live with him. When my grandparents told me I was moving to Alaska, I was pretty excited. They told me that I was going to be living with my dad, but I didn't really know what that meant. I'd never really had a dad. It was kind of an alien concept for me.

Everything about the trip felt like an adventure. Packing up my things. Flying in an airplane with my grandparents. Looking out the window as we crossed the Gulf of Alaska.

"That's where your dad goes out to fish," my grandpa told me. I remember staring at all that water. It seemed endless.

My dad was waiting for us when we got off the plane. He looked nothing like my grandfather, who was as clean-cut as the day he got out of the Navy. My dad was a big man with wild hair and an even wilder beard. I couldn't believe this scary-looking guy was the person I'd heard so much about. *This* was my dad?

I didn't know it, but he was just as scared as I was.

⁓

There are different degrees of seasickness, from mild queasiness to crippling nausea. It affects everyone differently. Some people will even tell you it's psychological.

The lucky ones don't feel a thing unless the weather gets really rough. Others turn green as soon as the boat pulls away from the pier. It's like being in a foxhole. You can speculate all you want, but there's no telling how a person will react until he or she gets out on the water.

Seasickness typically goes away after the body adjusts to being out at sea for a while. Some people deal with it by staying in the fresh air and keeping their eye on the horizon. Others ride it out while resting in their bunks.

And then there are the pukers.

There are those who lose their lunch and immediately feel better afterward. There are those who can't hold anything down for a couple days and gradually get better. And then there are the projectile vomiters, who spew everything in their system, and I do mean *everything*.

That was me at four years old, my first time out on the *New Venture*.

I was up in the wheelhouse when it started. All I wanted to do was run around and explore every nook and cranny of the boat, but my dad wanted me stay where he could keep an eye on me—and with good reason.

I was okay at first. As the *New Venture* jogged out of the harbor, I felt fine. But as soon as we started making way in open water, the boat began to pitch up and down and roll from side to side. This was a brand-new experience for me, and I started to feel a little funny. It didn't take long for that funny feeling to turn into something unpleasant.

I think my dad knew what was going to happen before I did. He sent me down below to the bathroom, but I didn't make it. I puked all the way down the wheelhouse stairs. I must have hit every step.

With my grandma, Barbara Campbell, in 1978.

After I'd tossed my cookies, I went to my dad's stateroom—the captain's quarters—to lie down. We were crossing the Gulf of Alaska to the Prince William Sound, which is about an eighteen-hour run. It was a miserable night that seemed to last forever, but the next day I felt a little bit better. I went out on the back deck to take in some of that cold, clean Alaska air. When we reached the protected waters of the sound, things calmed down and I drank a little grape juice.

Big mistake. Even though the water was calm, I wasn't done being sick; I threw up again. But the next day, I felt fine.

Believe it or not, my dad never got over his seasickness. Even after thirty-five years of commercial fishing, he got seriously sick for the first few weeks of every trip. And not just early in his career, but *every single time* he went fishing. If it had been me, I think I would have found another way to make a living, but once my dad set his mind to something, he was going to do it and nothing was going to stop him.

We spent that first summer salmon tendering in the sound southeast of Anchorage, with me doing little odd jobs on deck. The sound was full of hundreds of seiners, so called because of the type of net they used to catch the salmon. A seine net has weights at the bottom of the net and buoys on top so that it sits upright in the water. The fishermen dragged their nets to catch the salmon, cinched it shut like a drawstring purse, and hauled it on board with a crane. The fishermen then dumped all the salmon they had caught in their holds and dropped the nets in the water again. Once their holds were full of salmon, they'd bring the catch over to a tender. That's where we came in.

A tender is like a floating freight hauler. Seiners don't have refrigeration systems. Instead of wasting a bunch of time and fuel going back and forth to the canneries, where the fish might spoil before it can get processed, salmon fishermen sell their catch to tenders. The seiners come alongside the tender, pitch all their fish into a giant basket called a brailer,

Me, Mom, and Dad, family picture, 1978.

and use a crane to bring it over to the tender. Then we'd weigh the fish, record the weight, and pop the lever on the brailer, and the salmon would tumble into our hold. The tender's refrigeration system chilled the fish until it was time to take it to the canneries.

That first summer there were probably four hundred seiners out there in the sound, with one tender for every twenty boats. Half the tenders would be out buying fish from the seiners, and the other half would be on the move, going back and forth either to or from the canneries.

For commercial fishermen like my dad, tendering was like a working vacation, especially when compared to the demands of crab fishing. The hours were regular, and the work was easy. Many fishermen would bring their families on board the tenders with them, and the *New Venture* was no different. That's how I spent my first summer in Alaska—salmon tendering with my new mom and dad.

Of course I wanted to be a fisherman, too. My grandpa had taken me fishing up in the Blue Mountains east of the Walla Walla Valley, but I quickly learned that salmon fishing was a lot different. You don't just drop your hook in the water and hope for the best. Still, I went out there with my fishing pole, waiting for whatever would come along. I didn't have much luck, but I wasn't discouraged. Every day I saw thousands of fish get dumped into the *New Venture*'s hold. I knew my time would come.

Occasionally the other guys on the boat would come over and give me pointers. One time, while they were showing me the proper way to hold my pole, I felt something tug on the end of my line.

"Hey, you got something!" the fishermen shouted.

Sure enough I'd hooked a fish. I couldn't believe it. The guys cheered me on as I pulled on the line. I pulled and pulled for all I was worth. It didn't take long. The fish was close to the boat, and the guys helped me get it out of the water and over the rail. It wasn't nearly as big as some of the other fish I'd seen, but it was a hell of a lot bigger than the fish my grandpa and I pulled out of the river up in the mountains.

On the fishing vessel *New Venture,* the first boat I went on with my dad, tendering salmon at age four in 1978.

I couldn't wait to show my dad. I tried to pick it up, but it was too slippery. The only way I could manage it was to bear hug it. I walked up and down the decks showing everyone my fish. I wouldn't put it down for anything. It was my first salmon, and I wasn't about to let it go.

I didn't realize it at the time, but a fisherman on the boat had taken one of the smaller fish that had fallen out of the brailers and hooked it on to my line while I wasn't paying attention. Little did I know, I was reeling in a half-dead fish. Hell, I wouldn't have cared if I'd known anyway. I packed that fish around with me like it was my best friend. By the end of the day, I was as smelly and slimy as the salmon. Before my dad would let me back in the wheelhouse, he had to strip me down and hose me off.

With my cousin, Chris Campbell, fishing on the Buskin River in Kodiak, Alaska, at the age of five.

The very first fish I truly caught on my own wasn't a wild salmon, but a sculpin. It was an ugly-looking, bullheaded bottom feeder that looked like it was half-fish, half-alligator. Even though it wasn't very big, I was afraid of it and didn't want to get near it. I got my picture taken with my sculpin, but I wouldn't touch it.

Kodiak Island is located just south of where the Aleutian Islands meet the mainland. If you imagine the Aleutians as a long, curved blade that extends westward into the ocean, then Kodiak Island is its hilt. It's about one hundred miles long and anywhere from twenty to sixty miles wide with steep mountain faces, dense forests, and countless coves teeming with wildlife. It had everything I loved about the Walla Walla Valley, only closer to home. Instead of being in the backdrop, the wilderness was all

Fishing for salmon on the Buskin River in Kodiak.

around me: mountains, forests, lakes, streams, and, of course, the sea. It was all right there, right outside my front door.

Salmon fishing is the lifeblood of Kodiak, and it's been that way for over a hundred years. All my dad's friends were fishermen, my mom's friends were fishermen's wives, and my buddies were fishermen's kids. That's just the way it was in Kodiak in those days.

Actually, most folks didn't have family on the island. In a situation like that, you tend to bond with your friends more tightly than you would elsewhere because you're doing things with them that you'd normally be doing with your family. We celebrated holidays and birthdays and shared milestones together. It's a different kind of bond. In many ways the fishing community of Kodiak *was* our family.

It was a lot like a military community: When the fishermen left, it was up to those left behind to pick up the slack and run the show. Spouses stayed home and took care of the kids. The bond my mom had with the other women in the community was very strong. They relied on each other and helped each other out, because they were all going through the same thing together.

The kids bonded, too, probably because there weren't that many of us. In those days only about three or four thousand people lived on the island, and 90 percent of them were connected to commercial fishing. During the summer, when the canneries were going strong, the population would double with college kids working for tuition money. Then, when school started up and the weather turned cold, the population would thin out again.

Commercial fishing has always been a young man's game, which meant there were relatively few fishermen's kids on the island. Our school was pretty small, and there were only two or three buses for the whole community.

I hated those buses. It usually took around forty-five minutes for the bus to show, and I had to wait for it outside, in every kind of weather. That toughened me up.

If the bus couldn't make it, there was no school that day. But we didn't go home. We stayed out and played all day long. Parents weren't concerned about anyone taking their kids, because everyone knew every-one. It was an island. Where would they take us?

My family didn't have video games or cable TV, and neither did the other kids I grew up with in Kodiak. We made our own fun. That meant being outside all day, every day, regardless of the season.

We rode our bicycles in the summer, and in the winter we rode snow-mobiles. After school we'd go squirrel hunting in the woods. My dad had a little .22 magnum chipmunk rifle that was basically a long pistol with a stock on it, and that's what I used. In Alaska, if you could walk and carry a gun, you could hunt. My friends and I took full advantage. We weren't *allowed* to come home unless it was dark.

My favorite thing to do was tear around on my mini bike. For months I'd thumbed through the Sears & Roebuck catalog, imagining which one I'd buy. There were three models: one that didn't have shocks, one that had back shocks, and one that had front and back shocks. I got the one with the back shocks. It had a little 50cc motor, and I loved taking it out on the beach with my buddies. Of course with all that sand, it was only a matter of time before it locked up on me. That was the worst feeling, walking the bike home. When my dad got back from fishing, we'd tear the bike down and he'd show me how to get the sand out of the air filter and carburetor.

I didn't pay attention the first time. I just wanted my bike fixed. When it happened again, I made sure I paid close attention so I would be able to do it myself the next time instead of having to wait for him to get home, which could be weeks and weeks. That's when I started taking an interest in how things worked. My dad was always willing to teach me what he knew, but I wasn't always willing to listen, until I realized I'd have to do it myself while he was gone.

The only thing I didn't like about Kodiak was school. When I was in first grade, school administrators decided to hold me back a year. My mom and dad attributed it to my unusual situation with my biological mother and all the moving around I did as a youngster. They figured being in a traditional home would help me in school.

They were right. I was never held back again, but it was a constant struggle, and there was more to it than my irregular upbringing. I had a hard time in the classroom, especially with reading. I was a quick learner. If you told me how to do something, I'd pick it up and never forget it. But when it came down to just me and the books, I really struggled. It wasn't simply that I was a slow reader. My reading skills were poor by any standard. I wasn't able to comprehend things at the same rate as the other kids. That's not an excuse. That's just the way it was.

As for reading aloud, forget it. There were times I'd tell the teacher I didn't feel well and get a pass to go see the nurse, because I was too embarrassed to stand in front of the class and stammer my way through a poem or whatever other torture they had in store for us that day. I hated knowing there was something I couldn't do that came easy to the other kids. So I wouldn't do it. If I couldn't do it well, I didn't want any part of it.

Instead, I spent as much time as I could outside, even in the winter. Hell, *especially* during the winter. The days were so short it seemed a shame to waste them on school.

I come from a fishing family. A very conflicted fishing family.

My dad had two brothers: My Uncle Dan, who was older than my dad, and my Uncle Kevin who was younger. Even though they had all grown up on the family farm, they made their living on the sea. They were pretty tight, and sometimes they'd all go work on the same boat together. In most families three brothers on the same boat would be a recipe for disaster, but they made it work—for the most part.

By today's standards, the *New Venture* would be considered a small boat, but in 1978 it was just two years old and one of the bigger boats in the Kodiak fleet. The *New Venture* was a Bender boat, meaning it was built in a shipyard in Bender, Alabama, which produced a lot of boats in the 1970s. There weren't too many Bender boats in the Alaska fleet because you had to take them through the Panama Canal.

My father and my uncle, Kevin Campbell, on the Kodiak docks.

The *New Venture* was a blue and white boat with the wheelhouse forward instead of aft. It also had a stern ramp so that it could drag nets for shrimp and crab as well as salmon. The man who'd had the boat built wanted to outfit it for multiple fisheries, what you call a combination boat. While he was a successful salmon fisherman, he didn't have any experience with the shrimp and crab fisheries and wasn't able to make a go of it, so he sold the boat to my Uncle Dan, who brought my dad on.

My dad and Uncle Dan turned the *New Venture* into a highliner, a top boat, the best of the best. The top 5 percent producers of any given fishery are considered highliners. When you're on a highliner, you know it. Everyone from the captain to the bait boy takes extra pride in being the best. Potential crewmembers will wait on the dock for a chance to get a spot on the boat, and other captains will talk about your successes in the bars around town. Fishermen are some of the most competitive people you ever want to meet, so there's always plenty of talk.

An eighty-six-foot long fishing boat might not seem very big, but to a little kid it was gigantic. I'd seen boats down south on the Oregon coast, but nothing anywhere near this big. Somehow it felt even bigger on the inside with endless cabins and compartments to explore. I could go up to the wheelhouse, down to the galley, and below to the engine room. Each space was its own world. In my mind the *New Venture* was massive.

Every summer, I went tendering with my dad for about a month or so during the salmon season. It was the only time I ever got to see my dad on a day-to-day basis.

Once I got the hang of things, my dad started paying me for my time on the boat. I started out at five dollars a day, and each year I got a little raise. Eventually I worked my way up to about twenty dollars a day, which is pretty good money for a kid. That's how I bought my mini bike.

Even though the money was coming out of my dad's pocket, being a paid worker changed things. I wasn't just a kid goofing off anymore. If the weather was bad or the seas were rough, I'd try to hide out and stay inside, but my dad always found me.

"You need to be out there working. You're getting paid to do a job, and you gotta do it whether you like it or not. You can't just do it when you want to."

You'd think being the skipper's son, I'd have it easy, but that wasn't the case at all. If anything, he was harder on me than he was on the regular crew. That's how he was taught when he was a kid. When my dad was twelve years old working on the family farm, my grandpa had him up on a ladder picking fruit, in the shop, and in the fruit-packing plant. My grandpa didn't pay my dad as a way to cut him any slack. That's not the kind of man he was. My grandpa watched over him to make sure he wasn't screwing around, and my dad did the same with me.

My job was to keep the deck clear when the seiners came over and to clean it up after they left. Every time we dumped a load of salmon, there would be about fifteen to twenty fish flopping around on deck. I'd pick up all the fish that didn't make it down the hatch and throw them into the hold. I also scraped the slimy white jellyfish off the deck with the hose and blasted them over the side. The jellyfish would get caught up in the net with the salmon and make a mess. There's nothing slipperier than a deck full of jellyfish.

It wasn't much, but it was my job. With any vessel everyone has a very specific job to do, and that was mine. Eventually I came to understand that if I didn't do my job, someone would have to take time away from his job to do it for me. I never wanted to be the guy who slowed things down or made somebody else's job harder. So I worked extra hard to keep up. I wanted to hold my own weight, even if I was only a kid.

We had some fun times, too. One day some of the guys on the *New Venture* were taking the skiff out to another boat to get some beer. I wanted to go really bad. I'd been stuck on the boat for weeks, and I was itching for something else to do, but my mom didn't want me out on the skiff. Finally my dad said, "Go ahead."

I slid down the back ramp and climbed into the skiff, and we took off. It was a bit choppy, but we buzzed across the sound like it was nothing. My mom's brother Bob was driving the boat. He'd already had a few beers and was a bit tuned up. I don't know if he didn't see the wave or if he simply misjudged it, but it got us, we went airborne, and we all went flying out of the boat. Somehow Bob got ahold of me and kept me close. Another skiff came along, and we got the boat flipped right side up. Before too long we were zipping around again as if nothing had happened.

I was soaking wet and freezing cold. Even though it was summertime in Alaska, the temperature rarely rises above fifty-five degrees at that time of year, and the water was close to freezing.

My mom had seen the whole thing, and she was furious. She's not a physically imposing person. Next to my dad she's downright petite, but she let them have it. I didn't get to leave the boat again.

Looking back, it was a dangerous thing for a kid to be doing, but there wasn't much in the way of rules up in Alaska in those days. People did whatever the hell they wanted to do and lived—if they were lucky— with the consequences.

Tendering in the protected waters of the sound was easy. Getting there and back, on the other hand, could be an adventure. One time, while crossing the Gulf of Alaska from Kodiak to Prince William Sound, the boat lost power. So much fuel had gotten into the oil pan, it blew the dipstick out. We started spurting oil and had to shut the engine down for a while so that my dad and the engineer could get the problem fixed.

The engineer was the number two guy on the boat, a man the captain relied on to keep the boat running. If something happened to the skipper and he couldn't do his job, the engineer would step in. But the engineer had some bad news for my dad.

"The engine won't start."

My dad and the engineer drained the oil, put new oil in, and tried to fire it up again, but it didn't work. They couldn't figure out what was wrong. In the meantime we were at sea with no power and no control over the boat, and we were getting tossed all over the place.

I didn't get seasick, but my mom freaked out, and that made me nervous. It was our first time being dead in the water. My dad kept saying we weren't in any danger, but my mom was certain we'd sink. It wasn't very rough, but when you can't control the stability of the boat, there's always the chance that you'll roll all the way over and capsize.

We had to call for a boat to come out and tow us back to Kodiak. We were out there in the Gulf of Alaska without any power for about twenty-four hours before help arrived. The boat was fine, but by the time we got back to Kodiak, our nerves were shot.

———

Every other year my grandparents would come up to Alaska for a visit, and on the years they didn't come, we went down to Milton-Freewater to see them. When we went south, I'd go spend some time with my biological mother. That was the deal she'd cut with my dad. So my summers were split between salmon tendering, visiting my grandparents, and seeing my biological mother. Though I enjoyed being with her, she didn't always want me around.

She wasn't like my dad, and that's exactly why I liked spending time with her. My dad and grandpa were cut from the same cloth. As far as they were concerned, if I wasn't working, I was wasting my time. As soon as I was big enough to handle an ax, I was splitting firewood and cutting up kindling. Picking up rotten fruit and clearing ditches at my grandpa's orchard. Sweeping up in my dad's shop. If there was a chore that needed to be done, I did it.

My biological mother was the exact opposite. She didn't care what I did as long as it didn't interfere with her partying. I could stay up late and watch TV or run around outside for as long as I wanted. She let me do whatever I wanted to do. I never got in trouble with her, because she was always drunk or high, and I never had to do any chores.

One time she gave me a six-pack, turned on HBO, and told me to have fun. I didn't see her again for two days.

When my dad or grandpa came to pick me up, it meant playtime was over. It was time to get back to work. I loved them, especially my grandpa, but they were hard-asses. Compared to them, I thought my biological mother was really cool. She was always partying and staying up late, and there were always plenty of people around. She moved around a lot, from apartment to apartment, house to house, so there were always new people to meet, new places to explore.

But as much as I liked visiting my biological mother, I disliked her boyfriends. Usually they were biker types or guys she brought home from the bar. I tried to stay out of their way, but it wasn't always easy.

One time I made a smart-ass comment to one of her coked-out boyfriends, and he just went crazy on me. He came at me with a knife and chased me around the house. I made a break for the door, but he kept coming after me. I ran around the house and into the neighbor's fields. It was wintertime, and I didn't have a jacket or any warm clothes on. I hid out in the orchard. I could hear him crashing around, screaming at me to come out. I stayed out there until it was dark. Finally, after I heard the motorcycle start up and drive off, my biological mother came out.

"He's gone! You can come back!"

I didn't believe her. I was scared out of my mind. When I couldn't take the cold anymore, I went back inside. She tried to make it up to me, promised me it would never happen again. But the next time she went out partying, she brought someone else home.

In the summer of 1981, we took a "vacation" to the coast. My biological mother and her flavor-of-the-week took me across the state to visit her dad and sisters. It was the first time I'd met my grandfather on my mother's side.

He was an ornery old fart. I said something he didn't like, and he backhanded me upside the head. I took off running and hid outside. He had a bunch of fishing gear in his yard, and I hid behind the net piles. I wouldn't come back in the house, which made him even angrier. Finally my biological mother coaxed me back in by letting me pick out a toy lighthouse to take home with me.

After the ordeal at her father's house, we went to visit her sister Mary. I'd never met her before. I thought it was going to be a brief visit, but shortly after we arrived, my mom and her boyfriend took off and left me with my Aunt Mary. There I was in a strange town, with someone I didn't know. A day passed, then two.

I tried not to worry. It wasn't like my biological mother hadn't done this to me before, but that was back in Milton-Freewater. This was different. I didn't know where I was, and if I didn't know where I was, how would my dad be able to find me? After a while I started to get really anxious. I would sit on the couch and watch television. I remember thinking, *By the time the show's over, she'll be back.* The show would end, and my mother would still be gone, so I'd watch another show and another show and another one after that. Each time I'd convince myself that she'd come back before the end of the show. I dreaded going to bed. After a couple days of this, I couldn't take it anymore and started to ask my aunt, "When's she coming back?" over and over again. I think I freaked her out a little bit. That's when she called my dad and told him what was going on.

When my dad came to get me the next day, I was relieved to see a familiar face, but upset that he was picking me up early when I was supposed to be spending time with my biological mother. It's easy to look back at it with clear perspective, but at the time it was confusing. I didn't understand what was going on, didn't know why my biological mother had left me at her sister's house for so long.

Later I found out that my Aunt Mary had told my dad that my biological mother had planned to kidnap and hide me where my dad wouldn't be able to find me, but instead of taking care of me, my biological mother ditched me with her sister and went on a bender in the bars. I don't think my Aunt Mary knew what her sister was really like until we came to stay with her, but once she found out, she realized I was better off in Alaska with my dad.

We didn't have a lot of money when we lived in Kodiak, but I didn't think we were poor. *Nobody* had a lot of money. Life was tough, especially in the winter, and we had some really lean times. We did what we had to do to survive.

If Alaska in the 1970s was another world, Kodiak Island was a whole other galaxy. Not only were we far away from the rest of the world, we were cut off from it, too. Anything that wasn't made in Kodiak had to be shipped there, and that drove prices up. So while you could buy a house for twenty thousand dollars and you were literally surrounded by wild salmon, everyday items were horribly expensive. It's still that way today.

I didn't know what real milk was when I was a kid. It was a luxury item in Kodiak. A gallon of milk cost four or five dollars. In today's money, that would be like twenty dollars a gallon. That was an awful lot to shell out for something that—because it took so long to get us—had a shelf life of only two or three days. So I pretty much grew up on powdered milk, and if we didn't have powdered milk, it was cereal and water. A big bulk bag of cereal with tap water was a pretty common breakfast. No sugar. No fruit. That's all I got, because that's all there was.

My main meal of the day was the free lunch at school. I ate as much as I could, because our suppers were sparse. With dad being gone, if we ran out of food and mom didn't have any money, we'd have to make do. That meant going outside to see what we could rustle up.

In Kodiak, hunting was serious business. We didn't do it for sport, but out of necessity. We lived in a house right on the water, and there were plenty of times when I'd have to go fishing after school so that we'd have something for dinner. We ate so much salmon, I got sick of it. My mom made it every way you can make it. We'd have salmon patties and salmon steaks and salmon this and salmon that. When we didn't have fresh salmon to cook, we always had dried salmon. Tons of it. To this day I don't really care for the taste of salmon.

During deer season we would take the *New Venture* out and go down to the south end of the island to hunt. At the time, you could shoot six deer per person. We'd take the whole crew and have at it. My dad would pick out a cove that looked promising and anchor up in the bay for a day of hunting. One guy would stay behind to watch the boat, and the rest would skiff in to shore.

Sometimes they'd leave me on the boat. They didn't want me in the woods, because I'd scare off the animals. I had to wait until the end of the hunt to go with my dad and try to shoot a deer for myself. It was a big deal for me, a boy whose father was gone so often. I treasured those days in the woods.

In the meantime, I got plenty of good practice at skinning deer. After the deer was dressed, we'd cure it out by hanging it in the rigging. We did this with all the deer, so by the end of the hunt, we'd have a pretty impressive display. If we had a half a dozen or so men aboard, and we each got our quota, we'd come back to Kodiak with thirty to forty deer hanging in the rigging.

Of course there was another reason why my dad did this: He wanted everyone in town to know we'd had a good trip. We weren't the only boat to go hunting on the island. In his mind he was in competition with all the other fishermen. Even though they weren't fishing, it was still a competition. Hunting, fishing, whatever—it didn't matter what they were doing. Everything was a competition to these guys. Whether it was thirty deer or three hundred thousand pounds of king crab, it seemed the only currency that really mattered on Kodiak was bragging rights.

Sure enough the other fishermen would be out on their boats, counting up the tally. It was a proud fleet. If a guy couldn't be the top king crab fisherman that year, he'd try to beat you at the next thing. That's how it was for the highliners. They always had a bull's eye on their backs. Everyone was gunning to beat the next guy. Because we were all fishermen's kids, it even trickled down at school. Someone would mouth off, "My dad caught more crab than your dad," and a fight would break out because they were jealous.

I remember when a third grader was picking on me after my dad caught more crab than his dad. I was just a first grader and new to the ways of Kodiak. When the final numbers for the king crab season were posted, his dad was pissed off that my dad beat him. So when his kid came to school the next day, he tried to beat me up, but he thought better of it when I refused to back down.

But back to putting food on the table . . . If we didn't get enough deer while we were out hunting, we'd have to go out on the road systems and poach deer to make sure we had enough food to get us through the winter. The funny thing is, we'd do it not in a big pickup truck like you'd think, but in this tiny '73 Volkswagen Bug that my dad had. It was all eaten up with rust, and there were holes in the floorboards. I loved to watch the road rush by. If the weather was bad and we had to push through snow, it would pile up around our feet like cotton candy.

My dad would park the VW Bug on the side of the road, and we'd go into the woods to see what we could find. One time we found a little doe and shot it. My dad had me run back to the road and stand lookout. I looked both ways and gave the all clear while he dragged the doe out of the woods. The trunk was in the front, because in a VW the engine is in the back. My dad threw the doe in the trunk, and we got out of there. Of course the trunk was rusted out, too, and that little doe leaked blood all over the road. Every time we stopped we'd leave a puddle under our feet. Any game warden could have followed the trail of blood in the snow back to our house.

It was pretty obvious during my childhood years that our finances were getting worse. The reason my dad had moved up to Kodiak Island in the first place was because the fishing was good and the processors were right there in town. He didn't have to travel far to catch crab, and he was making frequent trips back to Kodiak to deliver. He was in and out all the time. Sometimes he was in town only long enough for a meal and a good night's rest. That may not seem like a lot, but it was a hell of a lot

They start you hunting early in the Campbell clan. Here I am at thirteen years old with my first elk.

better than most fishermen had it. In those days the longest mom and I would go without seeing my dad was two weeks, three weeks tops.

But things changed. Each year more boats came out to fish around Kodiak. Our slice of the pie kept getting smaller and smaller. The crab stocks went way down, and the Alaska Department of Fish and Game closed up some of the fisheries.

The final straw came when I was eleven years old. My grandpa had some heart trouble and my dad was worried. But on this particular occasion when we got a call from my grandpa, it was about something else. They'd just gotten back from the hospital where they had received some bad news. My grandma had been diagnosed with cancer. With both of his parents sick, my dad decided it was time to go back to Milton-Freewater. There was only one problem: My mom didn't want to go.

I remember it very clearly. My dad had rented a big plywood crate for all of our furniture, appliances, and clothing—basically everything we owned. The crate would go into a container and be shipped to Oregon by sea. It was big enough that you could walk around inside of it. We had that crate backed up to the door so everything in the house could go right into it. The ship was sailing that afternoon, and we had to get it loaded or else miss the boat.

But my mom wouldn't help me load the crate.

Dad had gone up to Adak to go fishing, and it was up to me to make sure everything got loaded into the crate. I was still packing right up until the time the shipping company came to take the crate away. I was hustling around, and my mom was sitting on the couch. The shippers came, and the only thing left to pack was the couch she was sitting on.

"Mom, you have to get off the couch," I said.

"I'm not getting up."

"Mom, the shippers are here. You gotta move!"

That snapped her out of it. She got up, and I packed the couch in the crate. When everything was loaded, I nailed the door on the crate and off it went. After a tearful Christmas at a friend's house, we flew back to the mainland.

CHAPTER THREE

SKIPPER'S SON

After we moved back to Milton-Freewater, my dad bought the farm.

No, he didn't die. He bought an actual farm, which seemed like a weird thing to do. My dad was a fisherman. His brothers were fishermen. I thought the main reason they all left home and went out to sea was to get as far away from my grandpa's farm as possible. And my dad didn't buy just any farm, but the one right next to my grandpa's.

My dad knew our time with his parents was short, and he wanted me to take full advantage of it. So the main reason we moved back to the Walla Walla Valley was so that I could bond with my grandma and grandpa. What better way than to follow in the old man's footsteps as a farmer?

My dad would never admit it, but I think he was nervous about being able to make it as a fisherman. Some of the fisheries around Kodiak were drying up. The crab stock was way down. There was a lot of cod in the water, and they were eating up all the crab larvae. There was probably some overharvesting, too. So things were out of balance. They closed the Bairdi crab fishery for a number of years, and when they reopened it, the quota was very small. Then the king crab fishery closed in 1981, and it hasn't been open since. Things weren't looking good for the *New Venture*. My dad needed a fallback option, something he could rely on if he wasn't able to make a living fishing anymore.

Lots of fishermen were leaving the Gulf of Alaska, but they weren't giving up or going home. They were heading to the Bering Sea. Crab fishing was thriving there, but it was a whole different ball game. The weather on the Bering Sea was intense, especially during the winter. It made the Kodiak fisheries seem tropical by comparison.

In the Gulf of Alaska, a storm would last a few hours, a couple days at the most. That's how it works in the movies. The storm blows in, rages for a few hours, and moves on. When storm systems move over the Bering Sea, they can last for *weeks*. Winter storms make life hell for fishermen.

If my dad—with his farmland as backup—was going to make a go of it on the Bering Sea, he was going to need a bigger boat. Around that time, my dad got a call from my Uncle Kevin.

"Hey, I've got this boat," he said. "You want to come over with me?"

"Talk to me."

The boat was the *Arctic Lady*, and it was owned by a guy named Louie Lowenberg, who had been a partner with Uncle Kevin on another boat, the *Midnight Sun*. When Kevin moved over to the *Arctic Lady*, he brought my dad with him. Although the boat was based in Kodiak, the *Arctic Lady* did most of its fishing out of Dutch Harbor, smack dab in the middle of the Aleutian Islands.

That was a game changer.

Everything about the Bering Sea was bigger. Bigger weather and bigger waves. Bigger fisheries and bigger quotas. It also meant the seasons were longer than what my dad was used to in Kodiak—a lot longer.

It was a big move for my dad. One of the ways he and I are different is that he's always been a very conservative guy. When he moved over to the *Arctic Lady*, he was under a lot of pressure; the guy who ran the boat before him was a top producer, a highliner. If you take over for a top producer, you need to fish at a certain level or you'll be replaced by somebody who can meet the owner's high expectations. My dad knew going in that if he didn't produce, it was his ass. He was under tremendous pressure right from the get-go. He had been a highliner on the Kodiak fisheries, but could he cut it on the Bering Sea? Could he hack it in this desolate, dangerous place that was completely unfamiliar to him?

My Uncle Kevin, always pretty gung-ho, thought they could. Between the two of them, he was sure they could figure it out. And they did. But as far as mom and I were concerned, that was when everything changed.

Instead of seeing my dad every seven to ten days, now he was gone four or five months at a time. To me it seemed like he was gone forever. When he left, he stayed gone. From the time I was eleven years old, my dad was pretty much out of the picture.

Although I may have lost a dad, I gained a grandpa. In many ways my grandpa was more like a father to me than my real dad was. When my dad came home from fishing, there was always work to do on the farm. There were times when he'd step off the plane, and if there was a cold snap, he'd have to go out in the orchards to fill smudge pots all night long so the crop wouldn't freeze.

My dad worked all the time. That's all he did. His idea of downtime was to work on cars. It's just the way he was wired, but it was tough on our relationship. Even when he was home, I'd hang out next door and spend time with my grandpa instead of my dad.

I idolized my grandpa. It seemed like he knew everything there was to know about everything. He knew how to cultivate an orchard, work on cars, and do repairs around the house. When something broke, we didn't call a plumber or go to a mechanic. Grandpa fixed it. I thought there was nothing he couldn't do. Even though my grandpa was just as hard a worker as my dad, he'd mellowed with age and wasn't such a hard-ass anymore. He taught me how to play chess and would watch *Star Trek* with me after school. On weekends he took me up to the Blue Mountains and taught me how to hunt deer and fish in the streams. From the time I was eleven years old, we were pretty much inseparable.

Our farm was run down when my dad bought it, and he had to pour a bunch of money into it to keep it going. The previous owner grew prunes, but my dad converted the operation to apples and cherries. All the money he made from fishing got dumped into the farm. We didn't have enough money to hire people to do the work, so we did it all ourselves. And by "we" I mean "me."

Fairly small as farms go, just nine acres, our land contained seventy-five hundred trees. Maybe it's helpful not to think of it as a small farm but a really big yard with a shitload of trees, and I was the one responsible for the yard work.

When my dad was off fishing the Bering Sea, a lot of the day-to-day operation of the farm fell on me, especially as I got older. The more I could handle, the more I had to do. We had a couple farmhands, but only when we were planting or harvesting. The rest of the time it was up to me and my mom to take care of the farm when dad was gone. Every day after school I had to go out and work. It wasn't just my dad's farm, either. Depending on the season and what needed to be done, I'd also help my grandpa on his farm.

We had an old-fashioned, ditch-irrigated farm. Instead of using sprinklers and timers, the water flowed down channels throughout the orchard. A sprinkler system is more than just a ton of piping. You need a pump that runs all the time, which jacks up your electricity bill. We couldn't afford all that.

The problem with these ditches was that they were always getting clogged with weeds, dead leaves, and fallen fruit. A single plug could stop water from irrigating half the orchard, so clearing ditches was a constant task. The water had to get through, or the trees wouldn't produce or the fruit would be undersize and unsellable.

My job was to water the trees. On the surface it seemed like a simple task, but it was a lot more complicated than watering a lawn. I had to weed around the trees so they didn't get suffocated. I had to mow in between the orchard rows. I had to pick rocks and fruit out of the ditches to make sure they were completely clear so the water could flow. It was

My grandpa and me, when I was four.

like a fuel system: If you get contaminants in the line, the whole system breaks down.

I hated cleaning out those damn ditches. It seemed no matter how long or hard I worked, I could never get them all cleaned out. There were just too many trees.

The orchard was too big to tackle in a single day, so I'd work in rows. We had about one hundred rows. I'd water one set of twenty rows on Monday, then I'd change the weir, which was the watering system, and get the water going on the next set of rows on Tuesday, and so on. By the end of the week, the first set would be clogged again and I'd have to go in and start the process all over again.

It wasn't an ideal setup, because the trees were getting watered only once every five days, just barely enough to survive. So it's not like I could do it when I had time or if I felt like it. If I didn't do my job, those trees would die and there would be no fruit to harvest. It was a lot of responsibility for a kid still in middle school.

I was constantly plowing the ditches with a disk and hand shovel, turning up the earth to knock down the weeds and make it easier for the water to get through. But by making the ditch deeper, more rocks and weeds and leaves would fall in. It was a fight, every single day, just to get those trees watered. I was always behind. Change and rotate, change and rotate, always struggling to keep up. It was the first thing I thought about when I woke up in the morning. Hell, I'd even dream about it.

Sometimes a farmer above the ditch would take too much water, which came from an irrigation channel off the Walla Walla River, and there wouldn't be enough for us. Everyone was allotted a certain amount of water, but there was a pecking order. We were quite a ways down on the ditch. If somebody took more water than he or she was supposed to take, it meant there wasn't enough for us. Every once in a while, unscrupulous farmers would try to take advantage of the situation. They knew my dad was gone all the time. They knew it was just me out there. If someone was taking too much water,

I'd have to walk up the road, figure out who was doing it, and tell the farmer to knock it off.

These conversations never went well. I was just a kid. Why should anyone listen to me? But I never backed down. What choice did I have? I could either argue with these farmers or let them take our water and try explaining to my dad why all the trees died while he was fishing. That wasn't an option. So there I was, this little kid fighting with grown men over water rights.

Picking cherries at age thirteen on our farm in Milton-Freewater, Oregon.

One day an ornery old farmer didn't want to hear what I had to say. He cursed at me and chased me off his property with a shovel. If I hadn't run off, there's no telling what he would have done to me. I went down and told my grandpa, who was pretty crippled up by then with a busted back. But when he saw how upset I was, he jumped on his tractor with his shotgun and went up to have a talk with that farmer. My grandpa straightened him out, and I never had a problem with that guy again. That's how brutal the farming industry was in those days.

I suppose these experiences were good for me. They definitely toughened me up. Working on the farm provided plenty of lessons in accountability, responsibility, and the value of hard work. But the main thing I learned was that I didn't want to be a farmer when I grew up. I thought my dad and his brothers had the right idea when they left to go out to sea.

The toughest thing about working in the orchard all the time was staying behind while my buddies went up to the mountains or goofed off in town. I could never go with them, because I always had to work on the farm. I felt tied to the place. Even my friends who had farms could rely on their sprinkler systems and farmhands to help out with the operation. Nobody had to work like my mom and I did. Nobody. I'd come home from school and my mom would be out there with a shovel with blisters on her hands trying to keep the water down and the crop from drying up. Then it was my turn. I'd get off the school bus, throw my books in the house, take the shovel from my mom, and away I'd go. I'd be out there in the orchards until it was dark, longer if the weather was good.

On weekends it was even worse. I'd work eight- to twelve-hour days, trying to keep the farm going while my dad was fishing. For a long time I felt like one of those trees, just battling to survive.

It was what was expected of me. My dad likes to say that the only thing he ever gave me was the opportunity to work. In that case, he gave me a lot.

In many ways farming is a lot like fishing. It takes a lot of expensive gear and a ton of backbreaking work, and at the end of the day there are no guarantees. Sometimes you catch a lot of fish. Sometimes you don't. On the farm sometimes the crop comes in, but sometimes it doesn't. An early frost, a bad blight, or even a few weeks without water can ruin a crop. Both fishing and farming are high-risk, high-reward endeavors. It takes a rare combination of workaholic and riverboat gambler to make a go of it. You have to be willing to put in an insane amount of work at great cost and then risk losing it all in an instant. One good day won't make your season, but a single bad one can break you.

I took out all of my pent-up anger and frustration on my motorcycle. I had a Yamaha YZ80 that my dad had gotten me when I was twelve years old. I was as hard on that bike as my dad was on me. Maybe harder.

We lived alongside the irrigation channel and had to cross it to get into town. The Army Corps of Engineers had put in a dyke and laid a gravel road on top. Massive boulders were stacked on the downward side of the dyke to keep the river in channel. My favorite thing to do was ride my Yamaha screaming across that dyke. Like a lot of bikes, mine was loud. Even though the dyke was about a mile or so from the house, while my mom and dad sat at the kitchen table they could hear me as I went through all the gears and flew across at one hundred miles per hour. Not the smartest thing to do, especially on a gravel road. One wrong move and I'd slam into one of those boulders. That would have been all she wrote. But I was young, fearless, and pissed off—a combination that got me into plenty of scrapes. There wasn't much my parents could do but shake their heads.

One day I took off across the dyke and had the throttle pinned as far as it would go when the rod on the piston shot through the case. I nearly lost control. Somehow I was able to get it stopped without laying it down. I walked my bike back home, wondering what my parents were going to say. My mom stood at the door, shaking her head, while my dad helped me load it into the truck. We drove it over to the shop to see if it could be fixed. The mechanics shook their heads.

"We can't fix this. What the hell did you do to it?"

I told them I didn't know. There was a lot of head shaking that day.

My partner in crime was a kid named Justin Mason. We got into all kinds of trouble together. There was nothing we wouldn't try. If something seemed too sketchy, and common sense somehow got through to me, he'd be there to talk me right back into it again—and vice versa. We were a couple of hellions.

When I was thirteen my dad bought a four-wheeler. When he gave it to me, the only thing he said to me was, "Whatever you do, I don't want you riding on this thing with Justin. Is that clear?"

I nodded my head and went right over to Justin's house to show off our new toy. Well, you can guess what happened next.

We took the four-wheeler up to the river to see what it could do. There was this massive distribution culvert—a big concrete box—that was perfect for turning the four-wheeler around. I backed up onto the culvert, but I went a bit too far and the four-wheeler rolled off the back edge and flipped over with me and Justin still on it. There was a second culvert a few feet back, so instead of rolling into the river, the four-wheeler wedged in between the two boxes. It was a good thing, too. The four-wheeler didn't have any roll bars. If we hadn't gotten stuck there, Justin and I would have had about six hundred pounds of four-wheeler rolling over on top of us, and we weren't wearing helmets. We would have been crushed. To this day I don't know how we walked away from that one without a scratch.

Safety was the least of my concerns. I was a lot more worried about what my dad was going to do to me. The four-wheeler was stuck from rack to rack, and we couldn't get it out. I didn't want to go home to ask for help, so we went to Justin's house. After checking out the situation, Justin's dad took me home. I stayed back on the street while he knocked on the door.

"Now don't get excited, Scott," I heard him say. "The boys are okay."

My dad didn't say a word. We went back to the river and it took all four of us to get the four-wheeler out of there. It was completely totaled. I'd wrecked our brand-new four-wheeler the same day we'd gotten it.

I kept waiting for my dad to lay into me, but he kept his cool. We had a little shop at the farm where we fixed up the farm equipment that was always breaking down. My dad made me tear that four-wheeler apart. When I had all the busted-up pieces laid out on the floor, we repaired every single one. We straightened the racks and handlebars, sanded down the panels, replaced the broken parts. He told me what needed to be done, and I did the work. It took us a while, but we got that four-wheeler going again.

This is where I'm supposed to say, "I learned a valuable lesson," but that really wasn't the case. When it came to motorcycles, four-wheelers, and snowmobiles, I just never let up. I was constantly pushing the limits, and that usually meant crashing and wrecking my machines.

Even though I had plenty of accidents, I never broke any bones. All my buddies were breaking their arms falling off skateboards or breaking their legs laying down their bikes. I'd get banged up a little, but nothing serious. No hospital visits. That made me more than a little fearless, gave me a sense that I was different from other kids. Most kids my age were fearful, but I never went halfway in anything. For me it was all or nothing. That was my mentality then, and that's my mind-set now.

⚊⚊

Not everything changed when we moved back to Walla Walla. I still went to Alaska during the summer, the only real time I got to spend with my dad. Most summers, I went salmon tendering but when I was fifteen years old, I went crab fishing with him for the first time.

Due to the huge quotas, the opilio snow crab season was a lot longer back then and lasted all the way until the end of June. I took my tests early so I could leave school in May and fly up to Alaska to meet my dad for the last part of the season.

That first summer I met the boat on St. Paul Island in the middle of the Bering Sea. During the three-hour flight out of Anchorage, I looked out the window of the plane below and all I could see was the bright blue water of the Bering Sea. It felt like we were flying off the edge of the world.

I wasn't really looking forward to going out on the boat. I hadn't seen my dad in months, and now I wasn't sure it was a good idea. I'd never been crab fishing before, just salmon tendering, so I didn't know what to expect.

I recognized the *Arctic Lady* right away. She was different from the *New Venture*. She was kind of a unique boat. For one thing she had unusual colors—dark brown and tan—the only one of its kind in the fleet. She also had a huge mast and a tall wheelhouse. She was just an unusual-looking boat. If you saw a boat that was red or black or blue, you'd have to guess which skipper it belonged to until you were practically on top of it. That wasn't the case with the *Arctic Lady*. You could see her coming from fourteen or fifteen miles away. Even at night, with the lights way up high on our wheelhouse, other fisherman could tell it was the *Arctic Lady*. Day or night, you knew we were coming.

But the color scheme wasn't her only distinguishing feature. The *Arctic Lady* was famous for the naked lady painted on the wheelhouse. Every time we went to the shipyard, which was every two years, my dad would hire an artist to come down and paint a new model that the crew had picked out of a girly magazine. That image earned the *Arctic Lady* the nickname the "Frigid Bitch." My dad will tell you it's because we'd fish through any kind of weather, but really it's because we had a naked lady on the wheelhouse. We were just like the sailing ships of old, when it was thought that having a naked woman on the bow would calm the seas.

Over time the image got bolder. At first our lady just showed a little cleavage. Then she'd reveal more each year—like a striptease. When the *Arctic Lady* came back from the shipyard, all the fishermen in Kodiak were anxious to see what we'd done with her.

Finally she went totally topless, which got us into a little bit of trouble. One time a woman with a carload of kids came down to the dock and chewed us out over it.

"That's not right! I'm going to call the cops!"

"That's art," my dad said.

"That's pornography!"

So we taped up the Frigid Bitch until the tourist drove off with her kids, who had a sense of humor about the whole thing, unlike their mother. We took a lot of flack from the other fishermen.

"Why are you covering that up? That's censorship!"

That's Alaska for you right there. The people there will tolerate just about anything—except an outsider telling them what to do.

My dad didn't waste any time showing me around the boat. He took me down to my bunk and told me where to put my gear. I was starting to unpack when he hollered at me to come up to the wheelhouse.

Now, after years of salmon tendering, I thought I knew my way around a commercial fishing vessel. I had a sense of what I was in for. I thought—and I can't believe I'm actually admitting this—it was going to be *easy*.

My dad wasn't having any of it. He tried to prepare me for how hard my job was going to be.

"Crabbing is *nothing* like tendering. It's a whole different animal. You'll see."

I nodded my head, but how hard could it be? I was only going to be here for a few weeks, and then the season would be over. Piece of cake.

My dad laid it out for me. "Salmon tendering has set hours, but when we go crab fishing we work around the clock."

If I'd heard it once, I'd heard it a hundred times. And I didn't really believe the old man this time. I thought he was exaggerating. I thought he was trying to scare me. Fishermen were all alike, I knew. They competed at *everything*—even at how hard they worked—and they weren't shy about telling you about it. I'd been hearing fishermen talk about the grueling hours and lack of sleep since I was four years old.

To hear a fisherman tell it, no one has worked harder in the history of hard work than a fisherman. Fishermen are masters of the art of

The *Arctic Lady*, the first boat I went crab fishing on with my dad and uncle.

one-upmanship. If you tell a fisherman how you nearly lost a crewmember to a fifty-foot wave, get ready to hear about the hundred-footer that almost capsized his boat. It's just the way we are. We've also been known to exaggerate here and there depending on how many times we've told the story, how late it is, and how many drinks we've had when we tell it. I'm not calling fishermen liars, but when we're telling a good story and our competitive juices are flowing, we've been known to stretch the truth a little bit.

When my dad told me that his crew worked around the clock, I thought he meant *occasionally*. I thought every now and then we'd have to put in a long stretch. I thought it would be the exception, not the norm. Fishermen being fishermen, I thought the exception was what defined the experience. Well, I thought wrong.

My dad is the most no-bullshit guy I've ever met. If a boat is ninety-nine feet, three quarters of an inch long, most fishermen would call it a hundred-foot boat. Not my dad. It would be ninety-nine feet, three quarters of an inch every time. What I'm saying is, when my dad told me to get ready to work harder then I'd ever worked in my life, I should have listened.

I don't know what my dad had told the members of his crew before I got there—probably nothing—but they weren't very friendly to me. The only one who seemed happy to see me was the greenhorn who worked in the bait station. That was my job now.

As soon as we left the harbor I realized we weren't in the Gulf of Alaska anymore. The waves seemed swollen, jagged, crazed compared to what I was used to. These waves weren't fucking around. These waves were *serious*.

When we got out to the crab grounds, we worked much faster and longer than I could have imagined. On my first day eight hours went by in a blur. Then another eight passed. Then the hours just started to jumble together, and I would forget the most basic things, like what day it was or why I thought it was good idea to go crab fishing on the Bering Sea.

On my first "day" we went for twenty-six hours straight.

Wow, I thought, *this is intense!*

We got a four-hour break, and then we did it all over again.

It was like that *every day*.

⁓

Crab fishing is really simple. You put the pot in the water. The crab crawls into the pot. You take the pot out of the water. There's not much more to it than that.

If you've ever seen a lobster trap, a crab pot is essentially the same thing but on a much larger scale. Each pot stands seven feet tall, weighs seven hundred pounds, and costs two thousand dollars a pop.

Of course we didn't drop the pots in the water willy-nilly. We laid them out in rows called strings. Each string had anywhere from twenty-five to forty pots.

As soon as we were done with one string, we were off to the next one. The pace was relentless. There was no room for error. The guys on deck were pros. They knew what they were doing.

Not me. I was the bait boy. My job was to prep the bait. It was the nastiest, dirtiest, most thankless job on the boat, which was why they gave it to me.

When I was back in Oregon, I imagined all the things I'd do as a crab fisherman, but that first summer on the Bering Sea I spent all my time prepping bait. When I say all my time, I mean I was in that bait station every minute of every day. I prepped bait in the morning. I prepped bait throughout the day. I prepped bait all night long. All I did was prep bait. It was a perpetual struggle to keep up.

For the guys on deck, the days took on a familiar rhythm of setting the pots, picking up the pots, and getting the pots ready again. But for the bait boy, there's no rhythm, just a never-ending grind.

When you're hauling gear, you've got a fresh pot coming up every two and a half minutes. That means as the bait boy, I had two and a half

Landing a pot full of opilio crab.

minutes to help land the pot, get the old bait out, and put the new bait in. Whatever time I had left over before the next pot came up was for getting more bait ready. Except I never was able to get ahead, because I was always behind. I didn't take breaks. I didn't have time to eat. I was running on little or no sleep. I was going, going, going. And at the end of the day, what did I get?

The opportunity to do it again.

On the good days, I felt like a robot. On the bad days, I felt like a robot that everybody yelled at.

I didn't have many good days.

Even though prepping bait was the worst job on the boat, it was still an important one, downright critical to our mission.

In fact that was the first thing I learned on the *Arctic Lady*—there's no such thing as an unimportant job on a fishing vessel. Whether you're steering through a storm, tuning up the engine in the harbor, or doing paint and preservation in the shipyard, it all contributes to the boat's safety and success.

With that in mind, hell, I had one of the most important jobs on the boat. Without bait, you don't catch any crab, and if you don't catch any crab, you don't make any money.

Crabs are voracious eaters. They never stop looking for food. They travel in packs called pods, and they're always on the move. If the tide is going the right way, a pod of king crab can cover four, five, six miles a day. That's why you can be on top of them one day, and they'll be gone the next. You have to spread your gear so you can keep track of which way they're headed.

They're constantly chasing feed. Once they find some food, they eat it up quick and then they're off to find some more. When they stop to rest, they pile on top of each other and form large balls for protection from larger predators. When they're on the move, it can look like a wave rolling across the bottom of the seafloor. They just come in and annihilate everything in their path.

When crabs find something to eat, they go after it with their pincers and claws, which make noise. Crabs have excellent hearing, so when other crabs hear the sound of all those claws clacking together, they know there's food in the area, and they come to investigate. When they all move toward the food at the same time, it's like a feeding frenzy. As a fisherman, that's what you want. You want as many crab piling into your pots as possible.

And who puts the food in the pots?

The bait boy, the guy I just made a case for as the most important person on the boat.

Of course the rest of the crew on the *Arctic Lady* didn't see it that way.

On an average trip we went through about ten thousand pounds of bait, mostly herring and sardines. That's a lot of little fish. The bait comes in forty-pound blocks of frozen fish stacked on pallets. I was responsible for getting the pallets into the bait freezer so they wouldn't spoil. Then, once we got out to the crab grounds, I had to take out whatever we needed as we went along. The problem was sometimes it was just as cold *outside* the freezer as it was on the inside. Fish don't thaw when the temperature is subzero.

Whether it was ready or not, the fish went into the bait chopper. The bait chopper was one of the more dangerous tools on the boat, basically a box of stainless steel knives that spun around really, really fast. The bait chopper could turn a block of fish into sushi in a matter of seconds. Those knives were super sharp and could cut through frozen fish like it was nothing. I was terrified of that damn thing.

Every boat I've ever worked on has a sign that says, Do not put your hands in the bait chopper, and every captain has a story about a greenhorn who got his hand mangled in the bait chopper. What happens is a block of frozen fish will get stuck in the blades, and an overeager greenhorn will try to push it through the chopper and end up contributing to the bait in a way he hadn't intended. The crab aren't picky. They'll eat anything.

I even heard a story about a guy who lost part of his foot in the chopper.

"His foot? Seriously?"

"Well," the captain told me, "the sign above the bait chopper didn't say anything about feet . . ."

After the bait chopper ground up the fish into chunks, I stuffed it into bait jars and screwed on the lids. The jars had holes in them so the scent could leak out. That's what lures the crab into the pot. Once I got the bait jars in there, I needed to give the crab something to snack on. So in addition to the jars, I added hanging bait—usually cod. After the crab crawled into the pot, they started snipping away at the cod with their claws. That caused the clicking sound that said, "Hey guys, there's plenty food over here!"

We usually put about thirty pounds of hanging bait in each pot. If your average cod weighs ten pounds, a big fish and a couple of medium-size ones will get the job done. I'd cut them down the back and up the gut to expose the meat and release more scent for the crab.

Grind, stuff, and cut. Grind, stuff, and cut. Over and over again. All day and all night I was up to my elbows in freezing fish heads, fish tails, and fish entrails.

In spite of all that work, I made a fraction of what everyone else got paid. On a fishing boat you don't get paid a salary or an hourly wage. You get paid a percentage of the profits. If there's no profit, there's no pay. That's why it's important to work for a skipper who knows what he's doing.

Everybody on the boat gets a share of the profits that reflects his or her experience and expertise. For instance, the engineer gets a larger share than a deckhand because of his extra responsibilities.

When you start out on a fishing boat, you don't make as much money as everyone else. You get half a share, half the wage the rest of the deckhands receive. The greenhorns are typically the hardest work-ing guys on a fishing vessel, getting all the shitty jobs that they don't know how to do yet and having to work twice as hard and waste tons of energy figuring out how to do stuff that an experienced fishermen can do in his sleep.

It might not seem fair, but that's the way it is and that's the way it's always been for as long as anyone can remember. Even in the book *Moby-Dick*, a greenhorn like Ishmael only gets a fraction of what Queequeg makes, because on a whale boat a harpooner is a hell of a lot more valuable than some guy who shows up on the docks looking for work. It's the same way on a commercial fishing boat. The more you know, the more you can do, and the more you can do, the more you get paid.

As you get more experienced, you can move up in the ranks—just like in the military. Once you can do everything on deck, you get a full percentage share like everybody else. And I mean *everything*. Not everyone learns at the same pace. So it can take the whole season, or a couple seasons, or even years for some guys to become full share. Once you're full share you're always full share.

Unlike the military, where the officers and enlisted personnel move up the ladder on different tracks, there's just one path to becoming a captain. Every captain was a bait boy once. You have to start at the very bottom. You have to know every piece of equipment and be prepared for every situation. Anyone can run a crab boat when the seas are smooth, the boat is running well, and the crab are piling into the pots. It's when things go south that determines a captain's true worth.

But it all starts with bait. That's the way my dad was taught, and that's how he taught me.

Officially there's nothing lower than a bait boy, but unofficially I was several notches below that. For one thing I wasn't even a half-share. Technically I wasn't even a greenhorn. I was the seasonal help, some kid my dad paid out of his own pocket.

To make matters worse, I was the skipper's son. That meant the rest of the crew had to be nice to me. Being the skipper's son is like being on a team where your dad is the coach: He rides you harder than everyone else to set an example to the rest of the team, but all your teammates think you're getting preferential treatment. It was a lose/lose situation.

So I didn't know what I was doing, I wasn't part of the crew, and I was the skipper's son. Strike one, two, and three.

If I'd been a crewmember on the *Arctic Lady* in those days I would have hated me, too.

CHAPTER FOUR

FARMER'S DAUGHTER

A FEW MONTHS BEFORE MY SIXTEENTH BIRTHDAY, I BOUGHT MY FIRST truck with the money I'd made on the *Arctic Lady*. It was a little red 1960 312 four-speed, four-wheel-drive pickup that my dad and I saw in the parking lot of a Dairy Queen. I hadn't seen many like it before, and still haven't to this day, which makes what happened to it such a shame.

We were looking for a fixer-upper, something my dad and I could work on together and still have ready when my sixteenth birthday rolled around. We were sitting outside at the DQ eating our dinner and talking about what kind of rig I ought to get when I noticed the truck with a FOR SALE sign in the back window. It was a total primer job, but my dad always liked that era of trucks, so we decided to go have a talk with the guy when he came out of the DQ. We must not have been paying attention, because he drove off and we had to jump in my dad's truck and follow him to his house. He was waiting for us in his driveway. He seemed a little agitated.

"What do you guys want?"

"We're interested in your truck."

"Well, why didn't you say so?!?" he said and shook our hands.

We looked over the truck, even though it was already dark, and listened as he described its history and maintenance issues. He was asking for a couple grand, which was just about what I'd made fishing up in Alaska. I figured we'd slap a coat of paint on it and we'd be good to go. Well, I was wrong.

The next morning after the sun came up, it looked like a totally different truck. It was covered in primer all right, but a lot of the filler was cracked. With the hood up and the sun shining we could see it was mess in there. The engine was smoking a little bit and needed a tune-up. So what started as a simple paint job turned into a three-month restoration. We pretty much had to strip everything down to bare metal and start over. In a way it was good, because I learned a lot about auto body repair and the mechanics of my rig, but it took all the money I had left over and a whole lot of time.

On my sixteenth birthday we went and got my license and my rig was ready to roll.

Toward the end of the summer, my grandpa got sick. My mom took him to the hospital and they admitted him. It didn't seem like that big of a deal at first. My grandpa went to the hospital all the time, and this time didn't seem any different. Even when my mom told me that I needed to go see him, it didn't feel like a serious situation. It didn't really sink in that he was in a bad way. I was out doing chores when my mom came back from the hospital, and I told her I'd pay him a visit when I finished.

Instead of going straight to the hospital, I went over to a friend's house for a few hours and then came home to get something to eat. I was about to leave when my mom pulled into the driveway. I figured she had the same idea. I thought maybe we could go to the hospital together.

"Hey," I said. "I was just getting ready to leave."

"It's too late."

"What do you mean?"

"Your grandpa passed, Scotty."

I didn't know what to say. I couldn't believe it. My grandpa was gone? That just wasn't possible. I jumped in my truck and tore down the street. No seatbelt or nothing. I headed for the hospital, but partway there it seemed like a dumb idea. Like my mom said, it was too late for that. My grandpa was dead, and I didn't get to say good-bye to him. I pounded the steering wheel with my fists and stomped on the gas pedal. My anger was suffocating. I couldn't get out from under it. How could I have been so stupid?

I headed out of town. I don't know where I thought I was going. I was just going. Rocks from the gravel road rattled the undercarriage. Wheat fields blurred past my window. The needle on the speedometer crept past seventy and kept on going.

That old truck wasn't the easiest thing to drive. It didn't have any power steering. It was like a boat: Once I got going, it was hard to change directions, which is true of most things in life.

I had that old truck floored. The speedometer went to ninety. The last thing I remember is trying to take a corner, and then . . . nothing.

I came to in the hospital, the same hospital where my grandpa had died a few hours earlier. One Campbell left and another one rolled in.

My poor mom was a wreck. As if she didn't have enough on her plate. My grandma had passed away the previous year, and this was becoming an all-too-familiar scene for her. When my dad got the news about grandpa, he was halfway across the Gulf of Alaska. He was on his way to the shipyards, but still a couple days away. That meant my mom had to handle grandpa's passing and my hospitalization all by herself. I'm sure she had her doubts about us Campbells, but unlike my biological mother, she stood by me the entire time. That's just one of the million reasons why I consider her my real mom. She's always been there for me. Always.

They told me I'd flipped the rig and rolled seven or eight times. I was thrown out of the passenger window and landed in a dirt field. I ended up about three hundred yards away from my truck. They said the only thing that had saved me was that the field had been freshly plowed, which provided me with a nice soft landing. But it easily could have been my grave. After all my bitching about how much I hated farmwork, it was a farmer's plow that saved my life.

I didn't break any bones, but I was pretty messed up. I couldn't walk for a while. My back was badly bruised. As for my rig, it was completely totaled. At first my dad thought there was a chance we might be able to fix it, but once he laid his eyes on it, he knew it was too far gone. I'd had it for fewer than a hundred days, and I haven't seen another one like it since. It was a one-of-a-kind deal, like my grandpa.

———

When I started at McLoughlin High School, aka Mac-Hi, it was an opportunity for a fresh start. A clean slate. I was no longer the kid from Alaska who'd been held back a grade. I could be whomever I wanted to be.

I got into sports and discovered something else that was different about me: I was fast. Really, really fast.

I could run fast and I could run far. I was faster on a track and was quicker on a football field than anyone else in the school. No one could touch me. I was the fastest kid at Mac-Hi, and I made a name for myself as an athlete. But the thing about establishing a reputation is there's no shortage of people who want to challenge you for it.

I got my first test early.

The parking lot next to the track was reserved for seniors, and they were protective of their turf. After school the lot opened up and anyone could use it. That's where I'd park the truck my dad let me use for track practice. One time we went away for an overnight meet. I wasn't thinking and left my dad's truck in the lot. Big mistake. When I came back from the meet, his truck wasn't in the lot anymore. Some kids had taken it out. They just hooked up the back end of my rig and towed it out of there. Because I'd left the truck in park, it tore out the transmission.

Now there was something else for my dad to be pissed off about when he came back from fishing.

I asked around and found out who did it. His name was Aaron. I forget his last name, but he was a pretty tough kid. At least that's what people thought. He was a senior and I was a freshman, but I didn't care. The age difference didn't slow me down one bit. I went looking for him, and when I found him I confronted him with what felt like half the school watching.

At first he didn't know what to do. He was a senior with a tough-guy reputation, and here was this freshman getting in his face. That set him off, and we started to get into it—the usual pushing and shoving before all hell breaks loose. A teacher came out and broke it up. We made arrangements to fight later that night in a parking lot.

Most of the time these fights never happened. Someone usually backed down or didn't show. Or a parent or teacher would find out and break up the party. Cooler heads prevailed and the situation fizzled out.

Not this time.

I was still madder than hell. I'd spent the afternoon trying to fix my dad's truck, and it just made me angrier and angrier as the day wore on. My dad wasn't around to help me or talk me out of it or call up the kid's parents. He was a thousand miles away on the Bering Sea. Even if he had been home, I think he would have been just as pissed off as I was. You don't mess with a man's rig.

When I got to the parking lot, there were more than fifty people there. Aaron and his friends were surprised to see me. They thought I'd be too intimidated, too scared to show up. I was just a freshman after all. But when you've had to stand up to grown men all your life—fishermen, farmers, and bikers, men with shovels and knives and decades of hard work and pure meanness pounded into them—a high school punk with an inflated view of his own toughness wasn't about to change my feelings on the matter.

I didn't waste any time.

I stepped up. We fought. I beat him. It was a quick scrap that left no doubt in anyone's mind who'd been whipped and who'd done the whipping.

No one in the school messed with me again.

Well, that's not entirely true. There was one person who knocked me out on the regular.

Her name was Lisa.

Lisa was the prettiest girl in school. No, I take that back. Lisa was the prettiest girl in the Walla Walla Valley—or maybe the Tri-Cities. Hell, as far as I was concerned, she was and still is the most beautiful girl in the Pacific Northwest.

I'd seen Lisa around—she was impossible to miss—but we had gone to different middle schools. She came from a farming family farther out from town and went to the country school. Even though technically I was a farmer too, I went to the city school.

When I got to Mac-Hi, we were finally in the same building. We even had some classes together. We were both a little intimidated by each other. I was a hellcat and a troublemaker, which put her off a little bit. She was a total knockout and way out of my league. I didn't try to pursue her right away, but I admired her from afar.

It all changed during our sophomore year. I thought maybe there was a chance she might be interested. We'd been dating other people and going through the usual ups and downs of high school romance. In the spring of the second semester, we were both single; I started to put my game plan together.

Lisa blew me off at first, but I was persistent. I wasn't going to give up. Giving up is not in my DNA. But I wasn't getting anywhere. Every time I asked her out, she shot me down. I thought that deep down she liked me, but she was afraid of what others might say if she started dating someone who was good at playing sports and getting in fights and not much else.

Finally, on May 6—my birthday—I decided to risk everything and give it one last shot. I asked Lisa to go to the movies with me. I even tried to guilt trip her into it by making sure she knew it was my birthday. What kind of girl shoots a guy down on his birthday?

Well, Lisa did.

After that I left her alone. It was one of those rare moments in life where I "got the message." To be honest, I was kind of pissed. So I stopped pursuing her. Enough was enough. I wasn't going to waste any more time with this girl. I thought we had some chemistry, but I'd been wrong. It wouldn't be the first time, and it wouldn't be the last. Either way I backed off. I figured summer was coming and pretty soon I'd be up in Alaska anyway. Dating Lisa would only complicate things, so I put her out of my mind.

Of course that's when Lisa started to pursue me. The next week we went on our first date.

I still hadn't replaced the truck I'd wrecked after my grandpa died, so I had to drive my dad's yellow '78 Jeep Wagoneer to pick up Lisa. It had four-wheel-drive and was pretty much indestructible. It was a farm rig and looked and smelled like one. It was not the ideal vehicle for courting the prettiest girl in the Pacific Northwest.

The Wagoneer had a loose fan belt that made an awful squeaking sound. If I kept it above a certain RPM, the belt wouldn't squeak. Whenever I took the Wagoneer out, I was always rodding around town to keep the squeaking to a minimum. I didn't want to show up for my first date with Lisa in a squeaky junker that sounded like it was about to fall apart.

So when I pulled into Lisa's driveway, I came in a little hot. I was probably doing about twenty-five miles per hour. She had a gravel driveway, and when I hit the brakes I skidded sideways right up to the house, spraying gravel everywhere and pinging the siding.

Lisa's mom came out to see what the hell was going on. She called into the house, "Lisa, I sure hope that's not your new boyfriend."

When the dust settled, her dad came out, shook my hand, and took me out to the shop. Whenever my dad took me out to the shop it was usually because he had something to say that he didn't my mom to hear. A meeting in the shop meant bad news.

Lisa's dad wasn't much of a talker. He didn't say a word. He just handed me a rake and told me to get to it. Before I could take Lisa out, I had to rake the driveway where I'd tore up the gravel skidding into their lawn. I guess you could say it was a rocky start to a rocky relationship.

By the time Lisa and I started officially dating, it was almost the end of the school year. We had two weeks left until the spring semester was over, which sucked. I'd finally gotten the girl I'd been chasing all year, and now I was going to have to leave her. I was positive that by the time I came back at the end of the summer, she would be dating someone else. It had happened before. Girls I'd dated in seventh and eighth grade had dumped me while I was up tendering in Alaska.

It wasn't a very good feeling. In fact it was a terrible feeling, because it was exactly what my relationship with my biological mother had been like. She'd tell me she loved me and wanted to be a part of my life, and then she'd leave the first chance she got.

Why would Lisa be any different?

I went back up to the Bering Sea for another stint as a bait boy. This time I had a better idea of what to expect, but I'd be lying if I said it was any easier. Although I tried not to, I thought about Lisa—a lot. Back in those days there were no cell phones or Internet, and we had to correspond the old-fashioned way.

The mail service wasn't very good out in the Aleutian Islands. The fishermen were out to sea so often that the only time they'd get mail was at the beginning of a trip and at the end. When you want it the most is in the middle of the trip, when you're bone tired and homesick and questioning what you're doing so far away from home.

I was lucky I had my dad and uncle on board, but they sure as hell didn't give me any special treatment. When you're the skipper's son, the other crewmembers keep you at arm's length. They're not going to bitch and moan and goof around with you like they do with the other guys,

Lisa Scudder and I when we first started dating.

because they're afraid that what they say might get back to the captain. They know which side of the bread is buttered.

It made for a long, lonely six weeks on the Bering Sea. I would have been better off if I was just some guy who didn't know anyone on the boat. I couldn't wait to get back home, but I didn't want to get my hopes up too much in case Lisa had decided to dump me.

When we made our final trip into Dutch Harbor, there was a surprise waiting for me: a whole packet of letters. There were forty-two of them, and they were all from Lisa. She'd written me every day.

There's no such thing as a secret on a boat. The guys, seeing how excited I was, found my stash and read the letters aloud at the galley table while we were eating chow.

Oh, I really like you. I can't wait for you to get home!

The more embarrassed I became, the more the guys razzed me about it, but I was secretly overjoyed. Lisa had waited for me! Maybe she was different than the other girls. Maybe, just maybe, she was the one.

———

When I came home from Alaska, Lisa and I picked up right where we left off. A lot of other girls would have found another boyfriend for the summer, but not Lisa. She was the one who waited. If anything, being away all summer only intensified our feelings for each other. That sealed the deal as far as I was concerned.

I had everything I wanted. I had football. I had my friends. And now I had Lisa. I was the jock, the daredevil, and the clown. I was the guy everybody wanted to be. I was living the teenage version of the American dream—at least until my dad came home. Then it was the same thing all over again.

"How come you're doing so poorly in school?"

"What are you doing partying so much?"

"Why are you giving your mother such a hard time?"

"What the hell have you done to my apple trees?"

It's pretty obvious to me now that I had a major chip on my shoulder, that I was a kid with something to prove to the world. Whether it was on the playing field or on my motorcycle or partying with my friends, I was determined not to be outdone.

Unlike fishing, football was something I was actually good at. On the football field I felt like I could do whatever I wanted. I was tough, fast, and absolutely fearless. I loved playing football. I treated practice just like a game. When my teammates got gassed and started bitching about how hard it was, I'd think to myself, *This is nothing. This is so much easier than fishing.*

I was one of the best players on my team's defense. I loved flying around and hitting the ball carrier as hard as I could. I loved celebrating with my teammates after sacking the quarterback. I loved everything about it.

When it came to sports, my dad didn't give a damn. As far as he was concerned, I might as well have been playing cricket. He'd never been interested in football and barely understood the sport. He was gone all the time and didn't come to my games. To me it was a big deal when we won a game, especially if I'd played well, but my achievements on the field meant nothing to him. He thought it was all a waste of time.

I brought the same intensity and passion that I had for football to my partying. When I went to a party, I was the guy who was going to drink more and party harder than anyone else. Long after everyone had passed out, I'd be kicking my friends awake.

"Hey, wake up! I thought we were partying. What's going on here?"

No matter what it was, I went 110 percent. I still have that mind-set. When I do something, I'm going to do it better. I'm going to do more. I'm going to go above and beyond what others are willing to do, regardless of the risks.

There was just one place where I wasn't like that: school.

The only classes I was interested in were the ones I had with Lisa, and I still managed to get kicked out of those all the time. When my teachers confronted me about by bad grades and worse attendance, I'd just shrug my shoulders and tell them that I wasn't interested in school. But that wasn't the case at all. The truth was I was *terrified* of getting held back again and having to take special classes for remedial students. That just didn't fit with the image of the person I was trying to become.

A shitty student, I overcompensated in the things I liked to do: partying, playing football, and hanging out with Lisa, and I wasn't very discrete about it.

Looking back, I see someone who was sorely in need of an ass kicking. A few tried, but none of them were up for the job.

When I got into my senior year of high school, some of the younger guys tried to challenge me. They were usually a lot bigger than me, and some were stronger than me, too, but none of them were tougher than me.

They only knew how to act tough. They thought that because they were bigger the whole world ought to bow down to them. They thought that way because they'd never been challenged. They'd been in short scraps on the field or in the hallways at school or at parties where there were lots of people around to break things up. They thought that made them tough, but they didn't know what real toughness was until they fought me.

I was a scrapper. I'd been hardened by fishermen and farmers, knocked around by men who were older than me, and beat on by my mom's boyfriends. I wasn't afraid to fight. More important, I knew I could take a beating. If you could see me in action during some of my fights, you'd think I was getting my ass kicked—for the first minute or so. Then, when the other guy got winded, I got started. No matter what kind of punishment someone dished out, I knew I could take it, because I'd already endured worse. I wasn't afraid of anyone.

I think if my dad had been around and paid more attention and could see what I was up to, he would have set me straight. But that's the thing: He wasn't there.

It's not like my parents didn't try. They could see the path I was heading down, knew where it was going to end. More important, they knew that *I* couldn't see it because I was too young. I was a hot-headed kid who thought he knew everything there was to know in the world, i.e., a typical teenager.

My dad was always lecturing me about opportunities and windows and paths, but it all seemed so abstract to me, like he wasn't talking about real life. I didn't think any of it applied to me. It all went in one ear and out the other. I'd listen to him go on about this or that, and the whole time I'd be thinking, *You're never here. Why the hell should I listen to you? Say whatever you want to say. You're gonna be gone in two weeks, and I'm not going to see you again for months.*

I didn't give a damn what anyone thought, because I thought no one gave a damn about me.

It came down to respect. I was proud of what he did and what he'd accomplished as a fisherman, but the older I got the more that world felt closed off to me, like it didn't have anything to do with me. I hated how he acted like such a hard-ass all the time, and I hated that my mom put up with it. I sure as hell didn't respect my biological mother or the derelicts she ran with. I didn't respect my teachers or the administrators at school.

Bottom line: I didn't respect anyone.

The two people I respected most—my grandma and grandpa—were gone. Their deaths left me hollow. I filled that emptiness with hurt and anger. I was a time bomb, and the clock was ticking. It was only a matter of time before I exploded.

Whenever my dad had to leave for king crab season in the fall, he'd give me these speeches about what he expected from me while I was gone. King crab was the big one. That meant we weren't going to see him again until Christmas—if then.

It was the same old song and dance. "You gotta be more respectful to your mom. Stop partying so much. Stop driving like a maniac and acting stupid."

Now that I'm a dad, I understand where he was coming from. He was tired of getting calls from my mom. "Junior did this . . . Junior did that . . ."

But the thing he was most insistent about was school. He was always on me to do better in school. No one could understand why I did so poorly when Lisa was passing all her classes with flying colors.

"You gotta graduate," he told me.

If he said it once, he said it a thousand times. He kept pushing and pushing and pushing. Every time he pushed me to do better, all I heard was, "I'm not good enough," which just made me want to push back even more. His advice and encouragement—though well intended—had the opposite effect.

It was only a matter of time before all the anger and emotions came to the surface. Even then it seemed inevitable. We were headed for a major confrontation, but when the shit went down, it started with my mom.

I was downstairs in the basement with Lisa when my mom came down and caught us having sex on the couch. She'd come down to ask Lisa to help her make some cookies, and there I was with my pants down around my ankles.

"Scotty! What in the . . ."

She was absolutely mortified. I was supposed to be mowing the lawn, not messing around with Lisa. She sent me outside to finish cutting the grass while she got on the phone with Lisa's parents and told them what we'd been up to.

"You need to have a talk with your daughter!"

Lisa was embarrassed as hell and couldn't face my mom, so she came out to the shop with me. Well, one thing led to another, and we decided to finish up what we started. We were going at it when my mom walked in on us again.

Oh, man, was she pissed.

She had me take Lisa home. When I got back I mowed the lawn like I was supposed to be doing all along, while my mom fumed inside.

My dad was out fishing, so she had to wait a few days before she got him on the phone. But when she did, she didn't waste any time telling him what I'd done.

I think my dad was somewhat sympathetic. He remembered what it was like to be a teenager. Besides, he was hardly a role model for conventional family values. His advice to me was to get a room.

But my mom was upset, as upset as I'd ever seen her, and my dad was sick and tired of getting calls about the trouble I'd gotten into. So when he got home, he wanted me to apologize to my mom.

I refused.

My dad was strangely reasonable about all this, but I knew there was a limit to his patience. "Don't make a big deal about this. Say you're sorry and we'll be done with it."

"No."

Now my dad got heated. "If you can't be respectful, you need to move out."

Even though I was still in high school, I was eighteen years old, and I thought being that old meant I could say and do whatever I wanted whenever I wanted. Like all eighteen-year-olds I was shortsighted, but I was also uncommonly stubborn. A terrible combination. I refused to back down.

My dad is a lot like me in that we both have short fuses, and that was enough to set him off. Of course it wasn't just this incident, but all the incidents leading up to it, all the back talk and surly glances, all the times I'd said one thing and done another, all the late nights and days I'd skipped school. It had all been building up to this.

So I stood up to him. I thought I was a pretty tough kid. Hell, at eighteen years old, I thought I was a pretty tough *man*. I'd been in a lot of fights. I'd show this old duffer a thing or two.

That was my plan anyway. They say everyone has a plan until he gets punched in the face, and that was the case with me.

I pushed. He shoved. I pushed back even harder, and my dad decided he'd had just about enough of my bullshit. He threw the first and only punch. I ducked just enough to catch it in my forehead instead of my face, but the blow knocked me back into the TV I had on the dresser and we both crashed to the ground.

He got me down on the floor and started choking me. I couldn't believe it. In all the fights I'd ever been in, no one had ever done this to me before. It was surprisingly effective. I didn't know what to do. I tried to get him off me, but he had me pinned. There was nowhere to go. There was nothing I could do. I was helpless.

Mom stood in the doorway yelling and screaming and telling my dad to get off me.

"If you don't stop, I'm calling the cops!"

That snapped my dad out of it. He caught himself and let me go. I went into the bathroom to clean myself up. I had a huge imprint from my dad's wedding ring on my forehead. It seems comical to me now, but back then not so much. I couldn't believe it. My old man had kicked my ass.

I went back to my room and started packing my things. I threw some clothes, my football gear, and my squirrel pistol into a duffel bag and left. I couldn't stay with Lisa. Her parents were just as strict as mine were, and there was no way they'd put up with that. Not even for a night.

I didn't go far. I went down the road a bit to my buddy's house.

His parents knew I was having trouble at home and said I could stay for a while, until things calmed down. The thing is, I had no interest in calming down. I wanted to be out of the house, out of school, out of Milton-Freewater.

I accomplished my first goal, but now I was stuck. With no car and no job my options were limited. Still, I was determined not to give in and go back home. That, in my mind, was the most important thing.

The days passed. A "little while" turned into a couple of weeks, and then those weeks added up to months.

My friend's parents sat me down and said, "Scotty, we'd love to help, but you need to go back home."

I said I was sorry. I told them they were right. But I didn't go back to my mom and dad's. I was never going back there again.

There was just one place left to go: my biological mother's place. She was living in an apartment on the other side of town. I wasn't my mom's only child. She had three boys with three different fathers, and one of her kids was staying with her, a kid named Chris.

I called her up and asked if I could come over.

After a long pause she said, "Something wrong?"

I told her I needed a place to crash for a little bit.

"Sure," she said. "Come on over."

When I went to live with my biological mom, she said that I could stay as long as I liked. Then we partied and the weeks went by.

I hardly ever went to school, but she didn't care. She didn't have a nine-to-five job. She'd get hired somewhere, and after a few months they'd let her go. It happened so many times she stopped looking for steady work. She went on welfare. She worked the system. She did whatever it took to get by. She did odd jobs here and there. She worked as a house cleaner. She cleaned bars—probably to pay off her tab.

I always thought she could do better for herself if she'd cut back on the partying. She partied just about every night of the week. It was the thing that made her who she was. She was the opposite of my dad, but she was just as extreme as he was. All he cared about was work. All she cared about was partying. Work and partying. Partying and work. I was caught between these two poles. I felt a great deal of pressure to choose. If I wanted acceptance, I had to be like them. But no matter what I did, the acceptance never came. In the end all they ever did was leave. The ultimate rejection.

One morning I woke up with a pretty gruesome hangover. It was a school day, but I wasn't thinking about going to class. School was the furthest thing from my mind, but my biological mother had to go to "work." She was getting ready to go clean some guy's house. Instead of showering, getting dressed, and eating breakfast like a normal person would, she was scraping up whatever drugs and alcohol were left from the night before, topping off the tanks before she went out and faced the world.

I remember looking at my mom and thinking, *That's me in twenty years.* Just like that I realized that I was going to end up like her if I didn't change. I'd have nothing. Be nothing. I'd be a party guy. I'd be like the dudes my biological mother brought home from the bar. It wasn't even a bad night, where she got all crazy or one of her boyfriends tried to beat me up. It was just a normal night of getting fucked up. A normal day of picking up the pieces.

That day a switch flipped in my head. It just clicked. *This isn't the life for me.* I finally understood what my dad meant with all his speeches. I was eighteen years old, about to flunk out of school, sleeping on a couch in an apartment where addicts rolled in and out 24/7. I needed a change.

That day I packed up my things. I didn't have much, but I couldn't find my squirrel pistol. It wasn't where I'd left it. My half-brother swore he hadn't touched it, but it was gone. Later on he told me what I'd suspected all along: My hunting pistol had been pawned for cash to buy drugs.

CHAPTER FIVE

THE COLONEL

IN THE SPRING OF 1993, I SWALLOWED MY PRIDE AND ASKED MY DAD if I could come home. I didn't even know if he was at the house or out fishing. I went by and there he was, working in the orchards. I didn't beat around the bush.

"I want to come home."

"Okay," he said, "but you have to abide by my rules until you're out of school."

Deal.

My dad's rules were pretty simple: respect my mom, obey my curfew, graduate from school. The first two were easy, but the last was going to be tough. It was the end of the first semester, and I'd pretty much flunked all my classes. While I'd been living with my biological mother, I'd been to school just a handful of days. I'd missed so much time, everyone thought I'd dropped out.

If I had any chance of graduating, I needed to pull straight As from that point forward so I could get Cs to graduate. I'd never gotten straight As in my life, but this time I was determined.

The first person I went to for help was Lisa. She was a good student and always trying to get me to take my schoolwork more seriously. There was a part of her that was attracted to my wild side, but mostly she just wanted me to be more responsible. As far as she was concerned, this was long overdue.

Next I went and talked to my counselor. She was surprised to see me. She asked me what I was doing there. I told her I wanted to graduate, and she looked at me like I was nuts.

"I don't know about that," she said.

"What do you mean?" I asked.

"You're so far behind."

"You're saying it's too late?"

"I'm saying you're not going to graduate."

I was stunned. I knew it was a possibility. While I was partying with my biological mother, I'd considered dropping out. But I'd changed. I meant it when I told Lisa and my parents that I was going to graduate, but when my counselor told me that I couldn't do it, it was *on*.

When somebody tells me I can't do something, I do everything in my power to prove him or her wrong.

I hit the books. I got tutoring. I even went to night school. My counselor was right: I was far behind, but once I started getting serious, my teachers noticed. They were pleased that I was finally applying myself, but they were skeptical, too. In their eyes it was a classic case of too little, too late. In order to graduate I was going to have to push myself harder than I'd ever pushed myself, and the man who pushed me the hardest was my physical education teacher, which was kind of a shock.

P.E. was my favorite subject. While I wasn't the best athlete at any particular sport, my speed made me competitive in all of them, especially track. So I was surprised when my P.E. teacher told me there was no way he was giving me an A. I was flabbergasted. I thought out of all my classes, P.E. was the least of my worries.

"But nobody can beat me!" I complained

"Right, but you *barely* win."

Whenever we ran a race, I'd start strong and leave everyone in the dust. When I was confident that no one could catch me, I'd start to ease up, coast along, do *just enough* to eke out a win.

"Winning is meaningless if you're not pushing yourself," he said. "You should push yourself every time you set foot on the track. That's why you're not getting an A in my class."

I thought he was picking on me, just another grown-up trying to be a hard-ass. But as my grades started to improve, it slowly started to sink in. Maybe my teachers weren't such hard-asses after all. Maybe they saw things in me that I didn't see. Maybe *that's* why they were so hard on me. They saw the classroom potential, but knew I wasn't giving 100 percent there. And if I wasn't giving 100 percent in P.E., which I actually cared about for shit's sake, what chance did I have in other subjects?

Then something happened that turned everything I thought I knew about myself upside down.

I was really struggling with my English homework and asked my teacher for extra help, which I'd never done before. All my life I'd avoided reading aloud in front of class, and now here I was doing it one-on-one with my teacher. The frustrating thing was that I wasn't getting better, even after class when no one was around. So my teacher sat me down and worked with me sentence by sentence, line by line, and I *still* couldn't do it.

She thought she knew what the problem was and arranged for me to take some tests. "Great," I thought, "here comes the short bus."

But the tests confirmed her suspicions. I had a learning disability. I was dyslexic.

Once my teachers knew what the problem was, they were able to work with me. The school had teachers who were trained in helping older students overcome this particular learning disability, and they were able to help. My teachers taught me how to process things better, how to comprehend things. Suddenly everything that had been impossible seemed very, very easy.

I got my grades up, and I was able graduate—barely. I had to work twice as hard as anybody else, because it took me twice as long to comprehend things, but I did it.

Too bad they couldn't have figured it out a little sooner.

After I graduated from high school, the *Arctic Lady* went into the ship-yards at Nichols Marine in Portland. I went with my dad to get the boat ready for crab season. We lived in trailers out by the yards, four guys to a room. We worked from dawn until dusk, ate a big meal, went to bed, and started over again in the morning.

Graduating from high school was a huge achievement, but it was just the first step. That summer I told my dad I wanted to be a fisherman. His answer was swift and unequivocal.

"No."

"What?"

"You're going to college."

I couldn't believe it. I'd been crab fishing with my dad the three previous summers and gone tendering with him since I was four years old, and now all of a sudden he didn't want me to become a fisherman? It didn't make sense.

My dad was encouraged by the way I'd applied myself at school and thought that I should go to college. It's not that he was against my following in his footsteps as a fisherman, but he thought if I had an opportunity to do something different, I should take it.

"Trust me," he told me. "You don't want this life."

I thought he was talking about the hardship of being away from home. Maybe he was trying to use my feelings for Lisa against me. But that's not what was on his mind. He was worried about something much worse than a broken heart.

The industry was changing, and my dad didn't like what he saw. King crab was an open-access fishery in those days, and the quotas were getting smaller and smaller. In a derby style of fishery, anyone with a boat and a commercial fishing license could come up to the Bering Sea and fish. When the total quota for the season was met, Alaska Department of Fish and Game would close the season and that was that. No more fishing.

During the season everyone was trying to catch as much crab as they could as fast as possible. There were no limits on the amount any one fisherman could catch. Three to four hundred boats would come up for the season, all competing against one another for every pound of crab.

With the quotas going down and the seasons shortening up, it was getting more dangerous out there. The competition was so fierce that guys were overloading their boats. At least two or three fishing vessels would go down a season. They'd just disappear.

Back then the commercial fleet didn't carry a lot of life-saving equipment. There were no requirements for life rafts or survival suits. The owners figured that if a deckhand went in the water, those things weren't going to save him anyway, so the boats didn't carry them. They were considered dead weight. That put the entire crew at risk when a boat went down. By the time the Coast Guard was able to reach the scene with their helicopters and cutters, it was too late. If a boat got in trouble, it went down; and if it went down, that was it. There were no survivors. The conditions were too severe to sustain human life for more than a few minutes. Every time a fisherman stepped on a plane to go fishing in Alaska, he didn't know if he was ever going to see his family again.

So when I started doing well in school, my dad got it in his head that I should go to college, get a regular job, and have a normal life—not a death-defying one.

But after working my ass off just to graduate, I had no interest in going back to school. College was for people like Lisa, who was studying to be a nurse at Walla Walla Community College.

I wanted to work with my dad and my uncle on the *Arctic Lady*. My dad, however, was dead set against it. He made it clear that there wasn't a spot on the boat for me.

That summer I went to work at a wood mill about an hour away in Pilot Rock. I moved into an apartment with Lisa. A lot of our friends lived in the same apartment complex, and we had a lot of fun together. We'd hang out after work and visit each other's places. It's what I always imagined living in a dorm would be like.

We did all kinds of things to save money and hardly ever went out. We made our own entertainment. Sometimes my friends and I would get carried away and build pyramids out of beer cans, which almost always ended up getting knocked down just as soon as we finished building them. Lisa wasn't too crazy about that.

I would make it up to her by cooking meals for her, which was something she didn't know I knew how to do. After being on my own so much, it was either learn how to cook or not eat. When I got back from the wood mill, I'd go straight to the kitchen to make dinner for us while Lisa studied for her exams. We didn't have a lot of money. Actually, we didn't have any money, but those were some of our happiest days. We thought we were poor, but as I would eventually find out, it takes money to have money problems.

I enjoyed the job at the wood mill, but it was dull and the work wasn't steady. I never knew how many hours I would get a week. I needed a challenge and I craved excitement. A lot of my friends from the football team had enlisted in the Marines, and they were always trying to get me to sign up. One day I went and talked to the recruiter in town, listened to what he had to say. He was very gung-ho to get me in the Corps. I told him that I had my heart set on being a fisherman, but if that didn't work out I'd enlist.

He kept calling me, trying to get me to come down to the recruiter's office, which wasn't easy since I was commuting to the wood mill most days. He made all these promises to me. He'd get me fixed as a diesel mechanic. Good duty station. Guaranteed pay. Whatever I wanted. I knew he would say anything to get me to sign up, but he was pretty convincing. I decided to give my dad one more chance.

"Dad, if you don't take me fishing with you, I'm joining the Marines."

My mom hated the idea, and that may have been what swayed my dad. Fishing was dangerous, but not as dangerous as the military.

"All right," he said. "You can come up this winter, but it's not going to be easy. You'll be a true greenhorn this time. If you can't hack it, I'll put you off the boat."

Here we go again, I thought. *More tough talk.*

I figured I owed it to the recruiter to tell him that I wouldn't be signing up, so I called him and told him my dad was giving me a shot on the boat.

The recruiter was pissed.

"I can't believe you're doing this," he said. "You're doing a disservice to yourself and to your country."

"Hey, wait a minute," I said. "I told you from the get-go I wanted to be a fisherman."

"If you go up to Alaska, you'll never amount to anything."

I had a few things to say about that, and our conversation got pretty heated, but I never forgot that guy's words.

—————

I saw a totally different side of my dad that season. I knew that fishermen called my dad "the Colonel" but I never knew why. That season I met the Colonel.

My dad hadn't changed his mind about wanting me on the boat. He was still against it and as adamant as ever. He figured the best way to get me to do something was to tell me I couldn't do it. So he used reverse psychology on me. You want to be a fisherman? Fine, be a fisherman. But I'm going to make you regret it.

He enlisted the aid of the crew. At the beginning of the trip, he took his brother—my Uncle Kevin—and all the deckhands aside and told them in no uncertain terms that he didn't want me on the boat.

"Don't give him any special treatment because he's my son. I want you to break him, make his life hell. Make sure he regrets coming out here and never wants to come back."

The crew was thrilled. Some of these guys knew me from the summers, some didn't. None of them liked me. They jumped at the chance.

We get to punish the skipper's son? It's on!

Of course, I didn't have a clue.

There was a method to my dad's madness. The reason why he wanted me to come fishing with him in the winter was because that's when the weather was the worst. The Bering Sea in the winter is a whole different animal. If you want to break someone, that's the time to do it.

What makes the Bering Sea unique is that you have all this deep water—two thousand fathoms of ice-cold water—that rises up to about one hundred fathoms in about a mile and a half. This is called the Bering Sea Shelf. Fishermen refer to it as "the Edge," because the edge of the shelf is where crab like to hang out. This huge underwater cliff creates an unusually strong current that pushes lots of feed into the area, making it an ideal habitat for crab. A strong current equals lots of feed, lots of feed means lots of crab.

However, the current is so strong that it can override the weather, and that's where you get the dangerous seas that make the Bering Sea infamous. Forty-, sixty-, eighty-foot waves are not at all uncommon along the edge—especially in the winter. That's where the crab were, so that's where the *Arctic Lady* was going.

We left Dutch Harbor in January 1995, and I noticed the difference right away. The seas weren't just rough, they were downright nasty. On top of all that, it was dark all the time and what daylight there was had to fight through a thick layer of clouds. Dutch Harbor gets two hundred days of rain a year, and in the winter it rains damn near every day. When we steamed out of Dutch Harbor, it was dark and cold and wet, and it stayed that way the entire season. We hardly ever saw the sun.

In spite of having spent parts of three summers out on the Bering Sea with the *Arctic Lady*, I got seasick right away. The swells were that intense. The person in charge on deck is usually the guy with the most experience, and on the *Arctic Lady* that was Dennis Helms, a native Alaskan from Kodiak. Everyone called him Bubba. I asked him if a storm was coming, and he just laughed at me. Even though I felt like shit, I tried to stay positive. It was my first full season as a member of the crew. I was only a half-share, but that was a hell of a lot better than what I'd been making as a part-time bait boy at the end of opilio season. This time I'd do it all: throw lines, sort crab, run the hydraulics, the full experience.

At the start of the day, whoever was running the boat would come down to the staterooms and wake everybody up. From that moment, we were expected to be up, dressed, and out on deck ready to work in twenty minutes or less. All on three or four hours' sleep. I was so tired. I just wanted to roll over and go back to bed, but I couldn't. I spent half the time trying to get upright and the other half in a mad scramble to get my gear on and grab something to eat. That meant most of the time I went out on deck hungry or in wet clothes.

The seasoned guys knew better. Before they went to bed, they'd laid out all the clothes they were going to wear and prepped some food to eat. When I got off the deck the night before, I was so tired that I just wanted to hit my rack. I wouldn't eat anything. I'd just tear off my clothes and go straight to sleep. The rest of the guys would eat something before they went to bed so they'd have energy when they woke up. By the time they got up, got dressed, and had something to eat, I was just barely crawling out of my rack.

What hurt the most were my hands. I was experiencing soreness that I'd never felt before. I'd worked my hands and wrists for so long that the tendons had gotten swollen, which caused my hands to clench up in my sleep. I didn't care why they were cramping, I just wanted to use my hands and they wouldn't open. They were all balled up like claws. What the hell? Was I turning into a crab?

The other guys saw me trying to get dressed with my deformed hands. "You've got crabber hands," they laughed.

I shook my hands, banged them on my bunk, and tried to pry them open and get them to work. They felt like they were asleep, just completely numb. I had to ask the guys to help me. My hands hurt so bad I couldn't even pull on my own boots.

I was always the last guy out on deck. I was hungry and lethargic and my clothes were damp. It seems like a small thing, but it was so dispiriting to start the day that way. I just didn't have the energy level the other guys had. I didn't have the seasoning.

As soon as I popped out on deck, my dad or my uncle would start razzing me over the loud hailer. Every morning I asked Dennis what he wanted me to do, and every morning he sent me to the bait station. It was all I was good for, prepping bait.

I'd break up blocks of frozen bait and grind it up in the bait chopper. When it came time to stick my hands in piles of freezing cold fish and fill up the plastic jars with two quarts of bait, my hands felt like they were on fire. They burned like someone was zapping my hands with electricity.

For the first few hours, my hands hurt so bad I could hardly do anything. The hardest thing was getting the lids on the bait jars. I'd try to spin them on without twisting my wrists, which was just impossible.

There are lots of things fishermen don't tell you when you're starting out, because most greenhorns are too proud to ask for help. And when they do ask for help, most fishermen would shoot them down, because they're cruel, sadistic sons of bitches.

I'd prep bait for as long as it took to get on station, and then it was time to start hauling gear. We usually pulled about fifty pots. That meant I had to have one hundred jars of bait ready to go. Two per pot. That would take four to six hours, depending on the weather, and then we'd have breakfast.

As soon as we got the string hauled, we'd turn around and set them right back into the water. I needed to have the bait ready to go before we started hauling gear. My job was to have that bait prepared before the end of the string so that we weren't wasting time. Because there were only twenty-four hours in the day, we could only haul so much gear. If I spent an hour making bait after hauling the string, that's a lost hour, which was unacceptable. I had to make sure the bait was done and ready to go.

Well, after a couple hours of sleep, no food, wet clothes, and crabber hands, I couldn't keep up. I don't think anyone could. The rest of the crew had to pick up my slack, because they knew the Colonel would come unglued if we fell behind. It was the crew's responsibility to prep the bait. The bait boy was at the bottom of the chain, but at the end of the day, we were all accountable. It was my job to make the bait, and if I couldn't keep up, they had to cover for me. They weren't happy about that, and to say they let me hear it would be an understatement.

They'd slam doors shut on me when I went to clean the crab out of the pots, and if I wasn't going fast enough, they'd whip the back of my legs with the door ties. That's how they broke you in those days. It was like being in the military and going through basic training—boot camp at sea. They didn't take it easy on me or cut me any slack. They hazed the shit out of me.

There's no cushy on-the-job training on a fishing boat. They throw you right in. Either you can hack it or you can't. During my first season they made me do everything the hard way. They didn't see any point in training me, because they were going to make things so hard on me there was no way I'd ever come back. As a greenhorn you have to prove that you're worth training before the crew will take the time and effort to train you. Why waste time on someone who's not going to make it?

The true test of a fisherman isn't that he makes it through his greenhorn season, it's whether or not he comes back for more.

By the time the first string was done, my hands would finally be loose enough to do my job. After breakfast we went right back out on deck again. We usually pulled three more strings—150 pots—with short breaks in between strings while the skipper ran the boat over to where we needed to go. The breaks weren't really breaks. I had to use the time to prep the bait for the next string.

An experienced deckhand can get ahead by prepping bait while hauling the string so that during the break he can go inside, change into dry clothes, and wolf down some food. I couldn't even keep up much less get ahead. While they were inside, they were rotating their gloves—swapping out wet gloves for dry ones and leaving the wet ones inside out so they could dry out. I didn't know to do that, so I was out there with wet gloves the whole time, and that's bad news.

If you don't swap out your gloves and keep your hands dry, the salt water will suck the moisture right out. So even though my hands were cold and wet all the time, I was actually drying them out, and one by one my fingertips started splitting. It was painful as hell. I was able to work through it, but as soon as I stopped working I really felt it. Even worse was when I started working again. Doing the simplest things —like tying a knot—was pure agony. I'd been working with my hands my whole life, but now they felt like they were betraying me.

That's the kind of stuff that makes you or breaks you. That's when the doubt creeps in. Is this for me? Can I take it? Why am I putting myself through this kind of abuse? Until you figure out the tricks of the trade, all you have are questions, and I didn't have any answers.

After those 150 pots were done, we had a dinner break. By then I was so wiped out, because I was doing everything the hard way. When you're hauling gear, you have to act fast and move quickly, but you have to go about it the right way so you don't burn any more energy than you have to. That comes with experience, and there's only one way to get it: by doing things the wrong way.

As a greenhorn I didn't even know how to walk correctly. With the wild weather and the way the boat was moving, I would waste all kinds of energy just trying to keep my balance. The boat would roll one way, and I'd overcompensate. Then I'd have to expend all this wasted energy just getting my balance back. It might not sound like much, but when you add all the wasted effort over the course of a twenty-hour day, it's a lot.

After dinner I had nothing left, but we still had one string to go, another fifty pots before turning in for the night. The problem was my body was already starting to shut down. My hands started to hurt again,

Me with my own crew and a pot full of opilio crab in the Bering Sea. This came years after my dad, "the Colonel," did everything he could to make me never want to fish the Bering Sea again.

my body was exhausted, and my energy was shot. I'd go back on deck dead to the world. That was the hardest time for me, the time when I most felt like giving up, and it took everything I had to keep going. I wanted to quit. Believe me, I wanted to get off that boat more than just about anything I've ever wanted in life.

The only reason I didn't quit was I didn't want to give my dad, my uncle, and the rest of the crew the damn satisfaction. They could sense that I was ready to give in because they'd seen it so many times before.

My dad and my uncle had a real aggressive style of fishing. Everything was predicated on speed and efficiency. The boat was fast and the crew had to be quick. The deck was arranged to make the process as simple as possible.

The goal was to hit twenty pots an hour, day after day after day. My dad was meticulous about it. He even used stopwatches. Everything was timed and recorded and evaluated for maximum production and consistency. That was his big thing, consistency. He didn't believe in kicking ass at the beginning of a trip or at the end or when he felt like it, but every single day. He demanded our best effort and refused to settle for anything less. That's how he got his nickname "the Colonel."

His style was to set the pots a half a mile apart and plow the bottom. He believed in long strings of 40 or 50 pots. Most guys would lay 20 to 25. When other crews were pulling 150 pots, we pulled 220 to 250 pots, and we did it consistently, so we were pulling more pots than the rest of the fleet every single day. At the end of the season, that difference was huge and resulted in up to a half a million more pounds of crab. More crab meant more money.

That's how my dad was able to push his crew so hard. They didn't want another boat to beat them. There was a sense of pride in being the best. The *Arctic Lady* was always in the top 5 percent. When the *Arctic Lady* came into port, all the other fishermen wanted to know how much she'd caught. It was the boat everybody wanted to beat and the standard by which other boats measured their success. There were plenty of good fishermen in the fleet. They'd have a monster season, and then the next year they wouldn't catch anything. They lacked consistency.

The Colonel was a pain in the ass to work for, but he was consistent. He was like clockwork: You knew what to expect. My dad knew he could push the crew to their absolute limits and beyond because they were making ridiculous money on the *Arctic Lady*. The crew understood that if they weren't the top boat, they were going to be close to it. For most guys it was worth the extra effort they had to put in because they knew that they were going to be rewarded at the end of the season. Guys could bank on making fifty thousand dollars per man during opilio season. Sometimes it was a little more, sometimes a little less. Other crews weren't making anywhere near that kind of money. They were making twenty-five thousand to thirty thousand dollars for the same season. The Colonel's crew would do whatever he asked them to do, no matter what. The guys lined up on the docks begging to get on the boat were a reminder that you could work your ass off and not make a dime if the guy in the wheelhouse didn't know what he was doing.

In 1990 my dad had to take the *Arctic Lady* down to Portland, where they cut it in half so they could put a 26-foot section in the middle. They lengthened the boat from 107 feet to 133. They put in new motors, new cranes, and raised the wheelhouse another 4 feet. A massive overhaul that went over budget by four hundred thousand dollars and cost almost a million dollars. You could almost build a boat from scratch for what it cost.

Naturally they ran behind schedule and had to do the sea trial on the Columbia River: a twelve-hour run from Portland to Astoria. They headed up to Alaska and hit a terrible storm, but the boat held up. They got to Dutch Harbor the day before king crab season opened. They got their gear on, went out, and caught $1,288,000 worth of crab in less than two weeks. They got on top of them and never lost them, which was difficult to do because king crab move fast. My dad and my Uncle Kevin were able to pay off two thirds of the loan in just two weeks. They put on over one hundred thousand dollars' worth of crab a day, and the crew made seventy-five thousand dollars each for twelve days of work. It was a pretty amazing feat considering the boat had been in two pieces just a few weeks before.

I'd heard the story so many times it was like a legend, and by the middle of the season it was the only thing that was keeping me going, the seventy-five-thousand-dollar carrot at the end of the stick. I was run ragged and close to the breaking point. No one would help me or show me how to do things. Fishermen were supposed to be like family. Brothers of the sea and all that bullshit, but these guys on the *Arctic Lady* weren't helpful at all. They went out of their way to make my life miserable, just as my dad and uncle had asked them to do. And when my dad or uncle was driving the boat, the other was riding my ass.

"What the matter, pussy? Are you ready to quit?"

That was hardest part: all that verbal abuse from my dad and uncle. But in a way that's what kept me going.

It was the last string of the season, the final fifty pots before we turned around and headed home. My uncle came down from the wheelhouse to tell me he wanted the pots baited up really good. Now, in my dazed and confused state of mind, I told him I would and went to work. Instead I stood there like a zombie. In my mind I was rocking it, slicing and dicing that fish like a sushi master, but in reality I was moving in slow motion. It took about fifteen seconds to cut a fish for hanging bait. I was taking much longer, like two minutes a fish, but I thought I was going like greased lightning.

"What are you doing?" my uncle screamed.

"What do you mean? I'm going as fast as I can!"

One of the crewmembers came over and pushed me out of the way and started cutting the fish the way it was supposed to be cut, and I realized how slow I had been going, because this guy was cutting five or six fish to every one of mine.

That's when I realized how burned out I was.

When we got back to Dutch Harbor, as mentally and physically broken as I was, I figured my dad had won. Of course I wasn't going to give him the satisfaction of knowing that he'd beaten me down, but I'd had enough.

After five weeks on the Bering Sea, I was ready to go home. I was done as a fisherman, and I was never coming back.

I knew it. The crew knew it. But my dad wanted to make sure.

When we got in the truck to go to the airport, I was so relieved. I had my bags in the back and all I could think was, *Thank God that's over!*

Dutch Harbor isn't very big, so when we drove past the turnoff for the airport, I knew something was wrong.

"Where we going?" I asked.

"I'm going home, but you're going cod fishing, son."

Before the season, when I was trying to convince my dad to take me fishing, I'd told another captain, a friend of my father, that I would go cod fishing with him on the *Beauty Bay* when crab season was over.

I'd totally forgotten about it, but my dad hadn't.

"No, way," I said. "I'm not going."

"Junior, you told him you're going. When a man gives his word to another man, it's just like a signed contract. All you have in life is your word, and if you can't honor your word, you're not a man. You can be a man and stay or go home like a pussy. Your choice."

I was squirming to get out, but there was nothing I could do. What was I going to say? I'm not a real man?

It was his best friend's boat. If I went back on my word, I would never hear the end of it. The Colonel had been saving this surprise for me all season. He was pretty sure that he'd beaten all of my desire to be a fisherman out of me but, just in case there was the slightest chance that I might consider coming back, he was going to deliver the final blow.

He took me to the *Beauty Bay*, had a few laughs with the captain, and went to the airport. He was heading home, but I had to spend the rest of the winter cod fishing.

The skipper told me it would be a thirty-day trip. Thirty days wasn't too long, but I knew that Lisa would be pissed. When my dad came home without me, she was going to freak out, but there was nothing I could do.

So I went longline fishing for cod on the *Beauty Bay*. With longlining, we used hooks instead of pots to catch the fish. It was an easier style than fishing for crab. It wasn't as physically demanding, but it took a lot more time. It was tedious, repetitive, and boring. Worst of all, there was hardly any money in it, which my dad knew, of course.

I'd never been longlining before, so I didn't know what to expect, but we didn't seem to be catching much cod. Thirty days turned into six weeks that stretched into three long excruciating months.

Toward the end of the trip I started getting sick. Not seasick. It wasn't nausea, but sharp stomach pain. I thought maybe I had eaten something bad. The skipper sent me to my stateroom, and later on he checked me out.

"Maybe it's your appendix."

"Yeah, maybe, I don't know."

The pain was intense. I tossed and turned all night long. The next morning it wasn't as bad. The pain was still there—a dull, agonizing pain—but not as bad as before. We only had a few days left in the trip. I figured since I'd made it that far, I might as well stick it out.

I finished off the trip, got on the plane, and went home to Milton-Freewater. The pain subsided considerably so that when I got home I'd pretty much put it out of my mind. I was just happy to be home and to see Lisa.

When I got there, my checks were waiting for me. The boys on the *Arctic Lady* had told me I was going to make some big money, but I didn't know what big money was. The most I'd ever made with my dad during the summer was two thousand dollars. I thought that was big money. Besides, I was just a half-share. How big could it be?

I opened the envelope and took out the check for the work I'd done on the *Arctic Lady*.

It was for thirty-two thousand dollars.

I opened up the check for the *Beauty Bay*.

It was for fourteen hundred dollars.

I didn't understand how one check could be so high when the other was so low. It made no sense to me. I thought there must have been some kind of mistake, but I didn't really care how it worked out. Thirty-three thousand dollars was thirty-three thousand dollars.

"Honey," I said to Lisa, "we need to go to the bank. Now."

That afternoon we headed over to visit Lisa's parents a few hours away. While we were there, the pain came back, and this time it was even more intense.

"What's the matter?" Lisa asked.

"I'm sorry," I said, "but I've got to get home."

We were two hours away from home when I started puking. We kept going, but we had to stop every five minutes because I was puking so much. I made it about halfway home before we went to the hospital. They ran a few tests and told me my appendix had burst. If I'd gone another four hours without medical attention, I would have died.

They admitted me to the hospital and took out what was left of my appendix. I didn't have any insurance. Even though my dad was on me about it all the time, health insurance was the furthest thing from mind. I was just a kid, barely out of high school. If my appendix had ruptured while I was fishing, the boat would have covered it, but since I was home, I was on my own.

My surgery cost twenty-eight thousand dollars. There I was, all excited that I'd made more than thirty-three thousand dollars, and all I had left over was about five grand. That money was my reward for working harder than I'd ever worked in my life and—just like that—it was gone.

Not-so-easy come, way-too-easy go.

When I was discharged from the hospital, they put me on bed rest until I was fully recovered. After a few days of sitting around the house, I started to get restless. Once you get used to working fishermen's hours, a regular eight-hour day hardly feels like work and no work at all makes you crazy. The worst thing about sitting around waiting to heal was all the time I had to think about the future. I was nineteen years old with no job and little money. The only thing I knew for certain was that I wasn't going to make any money with my ass sitting on the sofa.

I'd been out of the hospital for four or five days when my Uncle Kevin called me from Portland. The *Arctic Lady* was in the shipyards at Nichols Marine.

"Hey, we had a guy who didn't show up. You want his spot?"

"I just got out of the hospital," I told him. "I can't do much."

"Here's the deal. If you want a job, you need to be here tomorrow."

My uncle wasn't kidding. That's how hardcore he was. He'd drag a guy out of ICU if he thought he could get some work out of him.

"What's it going to be?" he demanded.

I had to make a decision. "I'll be there."

Lisa wasn't happy.

It would be more accurate to say that Lisa was livid. She was almost done with her nursing certification, and everyone she'd talked to told her that I should be taking it easy. She thought it was foolish of me to take a chance and risk getting hurt again. She thought I should stay home and recover. We talked it over, but there wasn't much to talk about. We needed the money.

"Besides," I told her, "it's not like I'm going fishing again. It's just for the summer."

Lisa was skeptical. She still hadn't forgiven me for being less than honest with her about how long I was going to be away. My previous trip had been really hard on her. She didn't completely believe me that I'd forgotten about my commitment to the *Beauty Bay*'s skipper, and she was still upset at me for being gone all winter and spring, when I'd told her that I'd only be away for a month.

Lisa wanted me home. We were young and in love, and my recovery had taken up most of our time together. Now here I was taking off again before I was properly healed. I told her I would only be in the shipyards for a month, which was technically the truth. But as soon as the *Arctic Lady* was done, the boat would be heading up to Alaska to go tendering, and I would be on board. But I couldn't tell her that. As much as I loved her, all I could see were dollar signs.

"Just for the month? You promise?" she asked.

"I promise."

Reluctantly, she let me go. I packed my bags and caught the next flight to Portland.

There are no easy jobs on a boat, but some are easier than others. When I got to the *Arctic Lady*, I took it slow and started out with some painting. That year we were changing out the windows in the wheelhouse. I was up on the scaffolding with my paint can and my brush, and I could see that my uncle needed an extra pair of hands. I went over to help out. It didn't seem that strenuous. I helped hold the windows in place while they were being bolted into the frame. Nothing to it. Someone down on the deck asked me something, and when I turned, I felt something give.

That can't be good, I thought.

Sure enough, I'd ripped out my stitches.

I had to go to the hospital to get stitched up again. It also meant another ridiculous hospital bill, because my dumb ass *still* didn't have health insurance.

There are all kinds of ways to learn a lesson, but I seem to be one of those people who can only learn them the hard way.

I went back to Milton-Freewater and celebrated my twentieth birthday with Lisa. It was a pretty low-key affair as parties go, but her most memorable gift came a few weeks later.

"I'm pregnant."

It wasn't a total shock—I'd proposed to Lisa shortly after I came home from the yards. I bought an engagement ring with what little money I was able to cobble together and left it on her pillow.

"So," I'd said, "you wanna get married?"

I'd be lying if I said I wasn't worried that she might say no. Lisa was well on her way to getting her nursing certification. All I had to show for myself was some nasty scars and huge hospital bills. But she didn't shoot me down, and we commenced celebrating.

We'd only been engaged a few weeks when the results of those celebrations became apparent. I had a hard time wrapping my head around it all.

Me? A father? It somehow didn't seem possible. Our wedding planning suddenly took on an urgency that caught me completely off guard, and the most unsettling thing about it all was that it wasn't just a ceremony we were planning, but the rest of our lives.

So I did what my dad would have done.

I went fishing.

CHAPTER SIX
DUTCH HARBOR NIGHTS

LISA AND I GOT MARRIED ON AUGUST 20, 1994. I WAS TENDERING ON the *Arctic Lady* up until about three days before my wedding. We had the wedding on the deck in my mom and dad's backyard.

Lisa had to plan it by herself—the ceremony, the reception, the honeymoon. Everything. I flew home, got married, went on our honeymoon, and went fishing the very next day.

When my dad got married, he went fishing the day after the wedding. And not just for a week or two, but for three months. Then I showed up. Just like that my mom went from being a new bride to a fisherman's wife and the mother of a four-year-old boy.

When you're a fisherman, there's no good time to get married. No good time to have a baby. No good time for anything. The boats work twelve months out of the year, and you never know when the season is going to end. Sometimes it's shorter. Sometimes it's longer. You just never know. It makes planning for major milestones next to impossible.

I was a twenty-year-old husband and father-to-be. A family man. I needed to make some money. I kept thinking about that thirty-two thousand dollar check and all the things I could have done with the money. I knew I'd feel a hell of a lot better about starting a family with some money in the bank. Plus, that thirty-two grand was a greenhorn's pay. As a full-share I'd make *twice* that much. Maybe a little more. Maybe a little less. There were no guarantees, of course. But fifty thousand to seventy thousand dollars for a few weeks of work—hellish, mind-erasing, body-destroying work—suddenly didn't seem so bad. Actually it started to look pretty good.

With Lisa pregnant I didn't have much of a choice. I had to take care of my family. That's when I decided I was ready to go crab fishing again. The money lured me back to the Bering Sea.

My father and I never really had a strong father-son relationship, but that changed when I told him I wanted to come back to the boat full-time. Even though my decision was made out of necessity, he understood where I was coming from. I don't think knowing he was about to become a grandfather had much to do with it. He'd done everything he could to break me, and now I was coming back for more. I'd finally earned his respect.

"If you're going to be a fisherman," he said, "I'm going to start teaching you the tricks of the trade. I'm going to show you how to be a good fisherman."

The Aleutian Islands are a fifteen-hundred-mile chain of mostly active volcanoes that stretches from Alaska to the Kamchatka Peninsula. The landscape is mostly tundra. There are almost no trees in the Aleutian Islands.

About ten thousand years ago, people started inhabiting these islands. They were good at fishing and not freezing to death. The people who inhabited these islands referred to themselves as Unangan, which means "the people."

In July 1741 Russian explorers arrived in the Aleutian Islands. They were led by a Danish seafarer named Vitus Bering. The first expedition worked out okay for Vitus, but the second was a disaster. He wrecked his ship and died after being stranded on one of the islands. The Bering Sea is named after this dumb ass.

Some of the sailors made it back to Russia with stories about the creatures they slaughtered to survive. They brought evidence of their escapades, including the luxuriously soft pelts of sea otters. This was bad news for the otters and even worse for the Unangan.

The United States purchased Alaska from the Russians in 1867. First it was a department, next it became a district, then a territory, and finally a state in 1959.

Everyone knows that Alaska is the largest state in the union, but the next three biggest states are California, Texas, and Montana. Alaska is larger than all three of those states *combined.*

The Aleutian Islands extend so far west that they cross the International Date Line. That makes Alaska the most northern, western, and eastern state in the union.

The Aleutian Islands were an irresistible target for the Japanese during War World II. They occupied the two westernmost islands, Attu and Kiska, and a Japanese naval force was dispatched to bomb the American navel base at Dutch Harbor on June 3, 1942—six months *after* the attack on Pearl Harbor.

After the military left, Dutch Harbor became kind of a forgotten place. It was so remote and far away that the only people who ever went there were fishermen. Alaska's state motto is "The Last Frontier," and that was certainly true of Dutch Harbor in the 1970s. There were no extradition laws, so if a guy got in trouble with authorities and needed a place to go where he'd have a clean slate, Dutch Harbor was the place. It was an unregulated town with no law enforcement whatsoever.

During the boom years of the seventies, Dutch Harbor was as close to a Gold Rush town as we'll probably ever see again. You had all these desperado types making a killing on the Bering Sea and coming back to Dutch Harbor with tons of cash and nothing to spend it on but booze and drugs—and there was a lot of it coming through Dutch Harbor in those days.

The drugs came from Asia. Rumor has it that dealers would take a trawler out like they were going on a regular fishing trip, meet up with suppliers from Asia, and do the transfer at sea. Then they'd come back into port and distribute the drugs down through Alaska to Seattle and the rest of the West Coast. I don't know how much of that is true, but there was a lot of cocaine in Dutch Harbor in the seventies. I've heard stories about skippers keeping big bowls of coke right in the middle of the galley table to help keep the crew awake.

Some boats were known as party boats. They would go out and fill up their tanks, and when they came back to offload, they'd party. They'd blow through four or five grand in a couple days and get back at it. They could do that because crab was a year-round fishery. They'd just go out and fill up their tanks again. It never occurred to them to put some of the money away. They weren't businessmen. They were just knuckleheads out for themselves.

Dutch Harbor was the last place in American society where a guy could stake his claim and make a million bucks by working harder than the next guy. That's what lured people to crab fishing in Alaska, and that's why they still come today.

For guys who wanted to work hard and play hard and didn't care about anything else, it was the perfect place to be. These guys didn't care about how much money they wasted on cocaine and alcohol. They just wanted to have a good time. In the Lower 48 they were considered hellions, the riff-raff of society, but in Dutch Harbor they fit right in.

It's not like that anymore. Crab fishing is big business. If there's money to be made, the corporations will regulate the ruffians right out. But there are still plenty of guys in the fisheries who are out of step with the rest of society. You have to be a little crazy to be a commercial fisherman. It takes a special kind of person to go out in a tiny boat and try to make a living on one of the world's most dangerous seas. Dutch Harbor is filled with those kinds of people, and that makes it an interesting place.

Now that the crew knew I was coming back, they were a bit more willing to show me the tricks of the trade. The fisherman who taught me the most was Bubba. Of all the people on the *Arctic Lady*—including my dad and my uncle—he had the most patience with me. He wasn't that big. He was actually a little guy. Looking at him you wouldn't think he was that strong, but he could outlast anyone in the fleet. He had more endurance than anyone I ever worked with. He could go for days and days and days. A robot. My dad would grind on him and grind on him, and he'd come back for more. He was the kind of guy who could work two days straight and afterward come up to the wheelhouse and volunteer to take a shift at the helm.

He taught me a lot. One of the ways I compensated for my learning disability in school was to be observant, a skill that paid off in the garage. Someone would show me how to replace a part, and I'd remember how to do it. But it was a different story on deck. Being left-handed, it was hard learning things simply by observing others. From throwing the hook to tying knots, the deck was set up for right-handed people. Bubba learned how to tie all the knots left-handed so he could teach me how to do it. That was just the type of guy he was. As long as I was willing to learn, he would take the time to show me what I needed to know to be an asset on deck. Bit by bit I learned what it took to be a fisherman.

In spite of all his knowledge and experience, Bubba was happy with his role on the boat. He wasn't interested in moving up the ladder. He had no ambition to get in the engine room or the wheelhouse. Bubba was happiest on deck. That was his show, and he didn't want anything else. He was perfectly satisfied with it. He wasn't the type to get jealous of others, which was good for me because I moved up the ranks a whole hell of a lot faster than I expected. Some guys didn't like it, but Bubba was always 100 percent in my corner.

At the end of the summer, the *Arctic Lady* was finally ready to go, but my dad and my uncle weren't going with it. Wade Vesser, the full-time engineer, was going to step up and run the boat. That left a vacancy. None of the other crewmembers were mechanically inclined, so I was picked to chief the boat, aka serve as engineer. This did not sit well with a lot of people.

On most boats the engineer is your second-in-command, the guy the captain relies on to keep the boat in tip-top condition at all times. I wasn't even a full-share, but as the engineer I got an extra percentage. It was unheard of for someone in his second season to be second-in-command. I wish I could tell you it was because I was exceptionally qualified but that would be a lie. I had no idea what I was doing on deck, but down in the engine room it was a different story, so I got thrown into the position.

Wade had skippered the *Arctic Lady* during the summer, but he'd never run a boat during crab season. Between me and Wade we had no experience at our respective positions. I don't blame the rest of the crew for being skeptical.

Knowing what to expect didn't make my second season any easier. If anything it was harder. Being the engineer didn't excuse me from my duties on deck, but it got me out of the bait station. I still had to be out there hauling gear and catching crab with the rest of the guys. I performed all my engineering duties in between strings or while we were on a run. So when everyone else was sleeping, I was transferring fuels, checking oils, doing whatever maintenance needed to be done. If I was able to keep up during the day, when we shut down I'd be able to do a quick walk-through and grab a couple hours of sleep before we went at it all over again.

Lucky for me the *Arctic Lady* was extremely well maintained. It went to the yards on a regular basis, and we knew what the boat had been through and what it could do. She had a big beautiful engine room that was twenty feet wide and twenty-five feet long and a nice tall overhead of twelve feet. It was a very deep boat that drew a lot of water, so there was plenty of space down there. I've been in boats where the engine room is tight and cramped and it's hard to get any work done. The *Arctic Lady* wasn't like that. She was a very comfortable boat.

For our main propulsion we had two main engines—Caterpillar 3412 V12 diesels—that ran at all times, twenty-four hours a day. Our generators were inline six-cylinder Caterpillar 3306s. That's what created electricity, ran hydraulics, and powered all of our refrigeration systems. If the generators weren't running, we didn't have any power. So if we lost a generator and couldn't get the other one up and running, the boat went dark. It was the engineer's job to get it going again. That meant going down to the engine room and fixing the problem. This happened more often than I would have liked.

Our biggest problem was seawater in the fuel system. Sometimes it would get into the tanks. Sometimes we got bad fuel from the canneries. Once the seawater got into the engine, it was only a matter of time before it clogged up and died. Didn't matter what we were doing—fishing, hauling gear, or traveling—it would just shut down. Whenever that happened I had to go racing down to the engine room to get the back-up generator going and figure out what the hell was going on.

We were always fighting the fuel. In the early days the boats didn't have centrifuge systems to clean the water out of the fuel. We used to deliver to a processor who had his fuel tanks underneath his refrigeration system. It seemed like every time we got fuel from that cannery it would come with a big slug of water. An engine won't run on water. If the engine shuts down, the generator won't run, so going dark was a pretty common problem in those days.

For whatever reason engine trouble always seemed to come during rack time. It never failed. I'd be in my bunk and hear the generator shift ever so slightly in pitch, which meant the generator was about to shut down. It may have been my first season as engineer, but I knew what the generator was supposed to sound like. I could be dead asleep, but the sound would wake me right up. I'd jump out of my rack and go flying down to the engine room before the generator died.

It was just a matter of being in tune with the boat. If the boat was pitching one way and all of a sudden it started pitching violently in a different direction, that told me something was wrong. If the water level in the tank wasn't where it was supposed to be, the boat would roll harder than normal. Maybe I would smell some antifreeze. A whiff of some fumes. A hot engine. A diesel smell that wasn't there before. These were the things a good engineer paid attention to. Anything could be a clue.

Bad weather could cause problems, too. If we hit a big storm and the boat was shaking real bad, it could mix up our fuel. That's when it got dangerous. The generators ran the pumps that kept water in the tanks. When those pumps stopped pumping, the water level in the tanks would go down. When the tanks were halfway full, I could feel the water sloshing around in there. That could threaten the stability of the boat. A boat with a full stack of gear, no power, and slack tanks was a boat just begging to roll over. Wade and I knew that, but it kept happening. I spent so much time in the engine room that it got so that I could maneuver around down there in the dark just as well as I could with the lights on.

It was kind of a spooky season, and not just because I was fishing without my dad for the first time in my life. The weather was fierce, and there were more boats in the water than usual. We'd just come off a strike, so guys were eager to fish. There were a lot of boats with inexperienced crews, and we were one of them.

That first night, we were about one hundred miles out on our way to the crab grounds when we ran into some weather. A cold winter storm with lots of freezing spray—perfect conditions for icing. A lot of boats got into trouble and *three* of them disappeared. One of them was the *Pacesetter*, a big old mud boat that was 140 feet long—a huge vessel by fishing standards. The thinking was that the bigger the boat, the safer you were. When word got out that the *Pacesetter* had sunk, everyone was shocked. The *Pacesetter* was considered unsinkable.

My idea of disasters at sea came from watching movies on TV where the ship goes down with great fanfare and the passengers get off in leisurely fashion like they're leaving a ball game. But it's not like that on the Bering Sea. When a fishing vessel goes down, it goes down fast, usually in less than a minute. It's not a long, drawn-out process. You're there one second, gone the next.

The Pacesetter was in front of us, and our skipper heard the doomed vessel's captain on the radio: "Mayday, Mayday, Mayday!" And that was it. No explanation of what went wrong. No tearful farewell. They just disappeared.

We were one of the first boats on the scene after the *Pacesetter* went down. We were driving along when all of a sudden we came upon some floating debris: a life ring, a buoy, a couple of plastic bait jars. That was it. There were no survivors.

Maybe they were overloaded with too much gear. Maybe a rogue wave hit them. Maybe they iced up and rolled over. No one will ever know. There are no gravestones on the Bering Sea. It was an eerie feeling to be out on the water knowing there were fishermen—guys just like us—taking their final rest one hundred fathoms below us on the bottom of the sea.

Every day I was reminded how dangerous the job was. I smashed my fingers. I bruised my ribs. I got smacked by a pot while I wasn't paying attention. I went too fast when I should have been going slow and vice versa. When the seas were rough, it was a constant battle just to keep my balance.

The most dangerous place to be was up on the stack, especially during a storm. There was no waiting for the weather to clear. If something had to get done, it got done. It didn't matter if the pots were stacked four or five high. Someone had to be up on the stack to secure the pots when the crane swung them over, and that somebody was me. I had to pay attention to a lot of things at once. I had to make sure I knew where I was on the stack so I wouldn't fall off. I had to keep track of the crane and make sure I didn't get hit with the pot when it swung my way. Most important, I had to keep an eye out for waves to make sure I didn't get swept over the side. It was a lonely feeling up there. Wind howling. Pots swaying. Danger all around.

I used to wonder, *If a wave takes me, will anyone even notice?*

Up on the stacks was the one place where my speed was an asset. I wasn't on a lanyard or tied up in any way. The pots were secure, but I wasn't. I had to be mobile so I could dodge the pots if they swung the wrong way or get off the stack if I saw a big wave coming. Sometimes I wouldn't see the wave, but I'd hear it. Every time a really big wave struck the boat, there'd be a half a second of dead silence. That meant we were in the comb of the wave. A solid wall of green water—no foam, no spray—blocked all the wind and wave noise and snuffed it right out. I'm talking total silence. When I "heard" that, I didn't even look up. I just grabbed on to whatever was nearby, because I knew it was too late to get out of the way.

A wave can crush you, kill you, sweep you over the side. When you get hit with green water, there's nothing to do but hold on for dear life.

I loved it. As crazy as it sounds, it was a huge thrill being up there on the stacks when the boat was bouncing all over the place, the wind was whipping at me, and the waves were crashing all around. The sound is incredible, and the water comes from every angle you can imagine: up, down, and sideways. Just an awesome feeling that's not like anything else I'd ever experienced. Nothing I'd ever felt on my motorcycle could even compare. Just pure elemental fury. That's when I knew I was a fisherman. The transformation was complete.

We finished up the season in November. It wasn't a great season, but pretty good considering our inexperience. I made about twenty-five thousand dollars and learned another painful lesson: Always pay your taxes.

Unlike a regular job, the fisheries didn't withhold any money for taxes, and it's up to the individual fishermen to pay the state and federal government what they owed. Not only did I have to pay money on the twenty-five thousand I'd just made, but I also had to pay on the thirty-three thousand dollars I'd made earlier that year.

My dad warned me not to make the mistake that most fishermen made and wait until the last minute. The next thing you know, the taxes

are due and you're out in the middle of the Bering Sea while the government is piling on interest and late fees.

I figured out what I owed. It was a huge chunk. I double- and triple-
checked my numbers, but the results didn't change: I was screwed. Between
taxes and my medical expenses, I didn't make any money that first year. Still,
we weren't starving. Back then our rent was $250 a month. My rig was paid
for. With rent and gas and groceries we were getting by on $500 a month,
which seems amazing to me now.

But all that would change when our baby girl arrived. We'd picked
some names that we liked: Kay Lisa, which was a combination of Lisa's
name and her mom's name; Ashley; and Stormee were the ones we liked
the most. What wasn't settled was whether or not I'd go back fishing
again.

Naturally, Lisa wanted me to stay.

Naturally, I wanted to go.

My dad was having a lot of trouble with the farm. He couldn't turn
a profit on it, and with me being gone and my mom helping Lisa, there
wasn't much he could do about it except go deeper in the hole. He needed
a good season or else he would go bankrupt. He was *counting* on me to
come fishing with him.

I was a full-share now. I could prep bait, sort crab, stack pots, throw
the hook, and run the hydraulics. I could do it all. If my dad and uncle
were the reason I refused to quit that first season, having them gone for
the second season gave me the space I needed to become a fisherman.
Still it was kind of strange to go from my dad doing everything he could
to break me to telling Lisa that he needed me on the boat. Talk about a
change of heart.

I didn't think I would be able to talk Lisa into it. I was just getting
started as a fisherman and already she was tired of my empty promises.
She'd listened to me when I was down and out and swearing that I'd never
go fishing again. She'd heard my promises of big money, but our bank
account was still empty. As far as she was concerned, all I did was lie.

There's some truth to that. I did mislead her, especially in the early days of my career. Instead of telling her that I wanted to be a fisherman, I told her I was only going to fish one more season. If we were going on a three-month trip, I'd tell her I'd only be gone a month. I kept making promises that I knew I wouldn't be able to keep. I wanted her to support me the way my mom supported my dad, but I was never able to come clean and tell her the truth. I was afraid that if she knew how long I'd be gone she would leave me.

Lisa was in her third trimester and heading down the home stretch. She wasn't going anywhere. She knew it and I knew it. So we both dug in until she finally relented at the last minute. I hopped on a plane and headed for Dutch Harbor.

———

With my dad and uncle back on the boat, Wade resumed his role as engineer and I went back on deck. Not as a greenhorn or bait boy, but a full-share deckhand. I didn't have the experience the other guys had, but they knew they could count on me. At the very least I was no longer a liability on deck. I'd proven myself, but the tests were far from over. We were all about to be tested.

The weather was exceptionally cold that year, and we started icing up before we even made it to the fishing grounds. That was always the most dangerous time to run into weather like that because we were carrying a full stack of gear. The moisture in the air clings to every surface and freezes. It gets into the pots and glues them together. Ice can turn a stack of pots into a mountain of ice and steel. If you don't do something about it, all that extra weight can compromise the boat, and by "compromise" I mean "sink."

It was so bad we had to anchor up behind St. Paul Island and deice. Deicing is a pleasant-sounding term for going out and knocking all the ice off the boat. It's not like a car, where you push a button and wait for the defroster to kick in. My deicer was a sledgehammer. We had to go out

Two examples of icing, one setting gear on the opilio grounds and the other a bow covered with ice when traveling in subzero-degree weather.

there and knock off the ice with brute force. If the sledge didn't work, we chipped at the ice with a crowbar. And if that didn't work, we busted out the jackhammer.

To get the ice off the bow, we lowered a pot over the side with the crane. I climbed out onto the pot and busted off the ice, which was eighteen inches thick in some places. We worked around the clock for two days. It was grueling, backbreaking work, like shoveling snow for forty-eight hours straight.

When we took off again, we went way, way up north, practically all the way to the Russian border. We were two days northwest of St. Paul Island when the ice built back up again. It was so bad the boat started to lay over on its side from all the extra weight. My dad tried to shift the fuel around to get us leveled out, but it didn't work. We were in serious trouble.

We were out over the edge, and the water was too deep to drop the pots. We didn't have enough line on them. We would have lost them all. My dad turned the boat around, and we started running for the edge. As soon as we got there, we started dumping the pots. We were tightlining them: dropping pots with one hundred fathoms of line in water that was one hundred fathoms deep. There was a chance we might not get some of them back, but we didn't care. We just wanted to get them off the boat as fast as possible, but it was slow going.

Each pot was seven feet tall and stacked four or five high on the deck. The pots on top had an extra foot of ice on top of them. I had to beat down through that ice to get to the pot ties so that we could get the hook through it. If I misjudged, I had to pick another spot and do it again. It was so cold that if I tried to cheat and break the pot free without knocking off the ice first, the pot's bars would break. It took five of us to knock the ice off the pot. Once we got it loose, we could move it off the stack, but our work wasn't done. We'd have to chip away with our sledges and crowbars to get the doors open. Then we'd have to break the lines and buoys free and get them deiced as well.

It was exhausting, time-consuming work, but we were in a life-or-death situation. We knew if we didn't get the extra weight off the boat, we could die. We were in survival mode.

Some of the pots had so much ice on them they actually floated. It messes with your mind a little bit to see a seven-by-seven-foot steel cage floating in the water. Some of them floated back over the edge and we lost them, but there was nothing we could do about it.

We worked as fast as we could. We were too tired to be scared, but up in the wheelhouse my dad was shitting bricks. Once we got forty pots in the water, the boat finally leveled out. It took us twelve hours to get those forty pots off, a process that usually takes only a couple hours. We worked harder than any of us had ever worked in our lives, and we hadn't even started fishing yet.

To make matters worse, my uncle's leg started acting up. He'd banged his knee and it started to swell. After a couple days, the swelling didn't die down. It kept getting worse and worse.

My dad and uncle were driving the boat in twelve-hour shifts. My uncle told my dad he wasn't feeling too good and needed a few extra hours of rack time. Knowing my uncle, he was probably in a ton of pain and didn't want to make a fuss about it. There are no sick days on a fishing vessel. If you can't suit up, you need to be dead or in the hospital, and there aren't any hospitals on the Bering Sea.

The next day my uncle wasn't feeling any better and needed a few more extra hours of sleep. The day after that he told my dad he wasn't able to come and drive the boat. So my dad ran the boat around the clock, pulling double duty in the wheelhouse. After a couple of days, my uncle came up and showed my dad his leg. He had a knot on the side of his knee the size of a small grapefruit, and it was an ugly shade of red like a balloon that was about to burst. My dad took one look at my uncle's knee and started steaming back to St. Paul Island to get him the medical attention he needed.

We were twenty-four hours out, and by the time we got to St. Paul's there were two issues: My uncle could no longer walk and the harbor was closed. There was so much ice in the water we had to just plow through it and hope for the best. We just pushed the boat in and got as close as we could. We had to put a pot in between the boat and the dock. We took my uncle off the boat with the crane and got him up to the clinic. The doctors examined his knee and knew they were out of their league. They medevac'd him to Anchorage, where they discovered he had a staph infection that had hollowed out the bone in his leg. If he'd banged it again or taken a fall his leg would have shattered, because there was hardly anything left to his femur. He had to get bone marrow transplants, and it took a long time to recover. He couldn't fish for eight or nine months.

It was unbelievably cold that year. Cold comes with the territory when you're fishing the Bering Sea. If you can't handle low temperatures, you're in the wrong place.

The Bering Sea has a way of throwing things at you that you've never seen before, putting you in situations you're not sure you can handle. The year I got married was the first time I ever saw ice fog, when the moisture in the air freezes and sticks to the boat. Everything we did made ice.

We'd haul a string and have to clear all the ice off the boat. Waves would wash over the deck and instantly freeze. Some of it would run off, but there was always some left over that would freeze and form a layer of ice on the deck. It would freeze up the scuppers, which are one-way valves that open up so the water can clear off the deck. They'd freeze shut, and the water wouldn't have anywhere to go. It couldn't clear and would form big blocks of ice.

It was so cold our crane froze. Our winches froze. Our hydraulics froze. It would take all five of us, scrambling around on an icy deck, to push the pots back. Absolutely miserable conditions.

Then a storm hit. A big, nasty winter storm.

Huge waves. Freakish winds. Massive seas. Just a cold, violent monster of a storm.

Keith Criner, aka Moose, using an electric jackhammer to remove ice from the bow of the *Seabrooke* while unloading cod at the Harbor Crown cannery in Dutch Harbor.

We were a couple days into it when it started blowing snow and the boat began to ice up again. It felt like the Bering Sea didn't want us there. I don't know how else to explain it. It felt personal, like it was trying to tell me something. I was thinking about Lisa and the baby, and I kept asking myself, *What am I doing here?*

We spent as much time deicing the boat as we did fishing, which meant we weren't making any money. You can work through the apocalypse when the fishing is good because you know you'll be rewarded for your work, but you don't get paid to drive around and knock ice off your boat all day. When I left home I thought I'd be able to make it back in time for the birth of my child, but with all the weather and ice and bad fishing, there was no way that was going to happen. It was completely out of the question.

Dad called over the loud hailer, "Son, you're the father of a seven-pound, seven-ounce baby girl. Congratulations!"

Suddenly the whole crew surrounded me. Everyone was cheering and slapping me on the back and shaking my hand. My dad was laughing on the loud hailer. It was like being in a movie. Just like that, I was a father. The world had changed. I was a new person. All the negative thoughts that had been swirling around in my head disappeared. It was like I'd been given a clean slate.

After about two minutes of high-fiving, I was back at it. Back to being cold and miserable and wishing I were anywhere but out on the Bering Sea in the middle of a winter storm. The storm didn't stop. Neither did we. We kept on doing what we'd come out to do.

When we finished the string, we had a thirty-minute run to our next set of pots, so I was able to go up to the wheelhouse to call Lisa on the phone system through single sideband radio. It wasn't a very good communication set up, but it's all we had at the time. I had to call a land station and get patched through. Usually I had to wait my turn, but not this time. I was able to talk to Lisa and see how she was doing. She told me about the baby. I told her about the storm.

"I guess that settles it," she said.

"What's that?"

"Stormee. We'll call the baby Stormee."

We didn't finish out the season until early April. I was a twenty-year-old father of a two-month-old baby. I made fifty thousand dollars that season.

The first thing I did was put a bid on a house across the state line in Walla Walla. I didn't have two years of tax returns yet, so I didn't qualify. I needed a cosigner. I asked my dad if he'd do it, and he said no. He thought the price of the house was too high and suggested a duplex in Milton-Freewater that my dad's uncle had built back in the day. His wife had gotten it in the divorce and was selling it. It made sense to keep it in the family.

We went and took a look at it, and I was sold. I put a down payment of twenty-five thousand on it, which was a dumb thing to do, because that was my tax money. But it was early in the year, and I figured I'd make more. My dad was nervous about signing because the apple crop was awful that year and he was close to being bankrupt on his farm. The last thing he needed was another debt hanging over him. I told him I was going to pay off the duplex in no time at all. It didn't quite work out that way.

That summer I went tendering on the *Arctic Lady*, continuing to work every chance I could get to pay off the mortgage. I told Lisa I would fish one more year and that would be it. This time I wasn't lying. I didn't think I could handle too many more seasons like that last one. I was a fisherman not an icebreaker.

Wade Vesser skippered the boat again, and I was the engineer. Fish and Game opened up the bairdi crab fishery, and we thought we'd give it a shot. No one on the boat had ever been bairdi fishing, but we figured it couldn't hurt to try, right?

Wrong.

We failed big time. Like opilios, bairdi is considered a "snow crab," but their habits are totally different. Bairdi are mud dwellers, and you won't find them anywhere else. Bairdi require a completely different style of fishing that was difficult to adapt to. You have to be patient and know what you're doing, and that wasn't us.

It was a horrible, horrible season. We weren't catching any crab and we weren't making any money. Instead of making back the money I'd used for the down payment, I went deeper in the hole. We fished through Christmas. We came home for five days and went right back at it for opilio season.

If I wanted to blow off some steam, I had to be sneaky about it. My dad and uncle didn't drink, so it was a dry boat. Alcohol wasn't permitted on board, and they expected their crews to stay sober at all times—even when we were in port. It wasn't like we had a lot of free time anyway. My dad didn't believe in hanging around town between trips, either. We'd come in, offload the crab, and head back out. So if I wanted to cut loose and have some fun in the bars, I had to do it during the twenty-four-hour window while we were offloading.

A typical workday lasted until around eleven o'clock at night, which was when my dad would go to bed. The older guys used the time in port to catch up on sleep, but the younger guys would go out and raise hell, and I was usually first in line. The deckhands would go out together. We'd party until two or three in the morning and sneak back on the boat. We had to be careful not to wake up the Colonel or else there would be hell to pay. Then we'd get up at six in the morning to go back to work again. We'd be hungover all day, but we knew we could recover during the twenty hours it took us to get back out to the crab grounds.

I think some of the guys thought it would be safe to go out drinking with me. They figured if the Colonel found out what we were up to, he wouldn't fire us on account of my being his son.

I wasn't so sure about that. My dad wasn't big on special privileges. The *Arctic Lady* was a top producer, and no one wanted to put his job in jeopardy. It wasn't worth losing a spot on a highliner for a few hours of fun in town. But I was always trying to see what I could get away with without pushing it too far. On some boats, partying was part of the deal, but not on ours. I liked to go out and have a good time every now and then, but I never really got out of control.

Well, almost never.

When we were working, partying was the farthest thing from my mind. I was 100 percent focused on fishing. But when we got back into town, I was like a caged dog that couldn't wait to bust loose.

I remember one three-day period in Dutch Harbor back when my dad and uncle were both running the *Arctic Lady*. After we finished up the season, my dad flew home while the rest of the crew got the boat ready for cod season. As soon as my dad's plane was in the air, Mike Bouray and I hit the bars.

Dutch Harbor was pretty much cleaned up by the time I came around, but I did experience one remnant of Dutch Harbor's sordid past: the Elbow Room. The Elbow Room was a tiny little bar in the village of Unalaska. To get there from Dutch Harbor you had to drive across a bridge, and for whatever reason it was always referred to as "the dark side." So whenever we went to the Elbow Room we'd say, "We're going over to the dark side, want to come?"

The Elbow Room was infamous. The old-timers would tell us stories about the knife fights, shoot-outs, and epic drug use that went down there during those endless winter nights. It was one of just three bars in Dutch Harbor. There was the UniSea Sports Bar next to the processing plant; Carl's, which was also on the dark side; and the Elbow Room. That's it. If we were drinking in Dutch Harbor, we'd find our way to all of them—usually in the same night.

Even though the Elbow Room wasn't nearly as rough as it had been during its glory days, it was anything but low key. For one thing it was small and always packed. It wasn't called the Elbow Room for nothing. With 375 crab boats in the fleet and six men to a boat, there was usually a lot of testosterone in the room. With the constant competition between the boats, you could always count on some kind of animosity among the crews. Scores to be settled, beefs to be hashed, and the usual bad behavior and reckless decision making that comes with excessive alcohol consumption.

But there were good times, too. If a boat had a good season, the skipper would ring the bell and buy a drink for everyone in the bar. It wasn't cheap. If there were 250 fishermen in the place at six dollars a rattle, that would set the skipper back fifteen hundred dollars. I can remember times when the bell would ring every hour because the captains were always trying to outdo each other. It was a status thing to come into the Elbow Room, ring the bell, and brag about how good your season was. If I wanted to know how everyone did that season, all I had to do was go to that little spot. The fishermen ringing the bell and buying drinks for everyone had a good season, and the guys who had a bad season were sulking in the corner. There was always a big party at the end of the season. Those were wild, wild times.

That particular night Mike and I started at the UniSea Sports Bar and then headed over to the dark side. Back then bars stayed open until four or five in the morning, especially at the end of the season, but for some reason we thought we needed some beer for the walk back to the boat. I don't know how we ended up with an eighteen-pack that late at night, or how we were going to get it back to the boat, but a police officer spotted us and pulled over her patrol car. She asked the same questions we were trying to figure out. The officer wanted to confiscate our beer, but Mike wouldn't cooperate. That's when I realized how absolutely shitfaced he was. The police in Dutch Harbor were used to dealing with drunken fisherman, but this lady didn't have much patience. I talked Mike into giving up our beer, and we went on our way.

We had a ways to go—all the taxis were long gone—and Mike could barely walk. Somehow we made it back, and while I was fumbling around on deck Mike climbed up on top of a pot. Before I knew it he was seven feet off the ground and had no idea where he was. It gave me a bad feeling seeing him up there.

I hollered at him to come down, but he wouldn't listen to me. I don't think he could even hear me.

"Don't move! Stay right there while I get some help!"

As soon as I turned to go, he walked off the pot and landed on his face. He was lying on the deck facedown in a puddle of his blood. I thought he was dead.

Mike busted up his nose pretty bad. Two black eyes. Scraped up lips. A real mess. But he was alive. It just goes to show there are no easy falls on a fishing boat—even when you're parked at the pier.

Of course I had to go wake up my uncle and tell him what happened.

Of course he was super pissed off.

Of course he blamed it all on me.

My uncle couldn't get mad at Mike because he was passed out drunk, so he took his anger out on me. "What the hell were you thinking?"

I didn't have an answer.

I got Mike inside and cleaned him up as best I could, but it was getting late. Mike's rack was right above Wade's, who was still the engineer on the boat, and he looked like he was thinking about taking Wade's bunk. I didn't think that was a good idea. Mike and Wade didn't get along. It was bad enough that Mike had woken everybody up. Wade wouldn't be too pleased to find Mike in his bunk.

"Hey man, you need to go to bed."

"Just a second," Mike said. He leaned forward and puked all over Wade's bunk. Mike cracked a smile through his bloodied face, crawled up into his own bunk, and passed out.

Needless to say, Wade wasn't too happy.

My uncle left the following day. The last thing he said was, "This better not happen again."

"It won't," I told him. "I swear."

Famous last words.

With my uncle gone, Wade was in charge. When the sun went down, I got the itch to go back out, and I convinced Wade to come with me. I had to twist his arm a little bit. I told him I would buy him a couple of drinks to make up for all the shenanigans from the night before. He gave in and we promptly got into some shenanigans of our own.

My dad always kept a beater truck in Dutch Harbor so that simple errands wouldn't take all day. We took the truck over to the dark side and hit all the bars. We may have hit the Elbow Room more than once. If so, it was one time too many.

On our way back to the boat, Wade failed to come to a complete stop at an intersection, and the telltale blue and red lights of a cruiser lit up the inside of the truck.

We pulled over and the police officer got out of the car. It was the same lady who'd taken my eighteen-pack the night before.

"Not this cop again!" I said.

The police officer didn't like that one bit. I thought maybe she didn't recognize me. Well, she recognized me all right, but it was what she'd heard that was the problem. When I said "cop," she heard something else, a single syllable word that starts with a C that is never a good word to use in front of any woman, much less a female police officer.

The cop cited Wade for driving while intoxicated. When she handed him the ticket, she said, "If it weren't for your buddy's big mouth I would have let you off."

This was bullshit, of course, but Wade believed her. To this day he still blames me for the ticket.

That was the end of the wild times on the *Arctic Lady*. After that, if someone got seriously hurt or arrested while drunk, they got fired. I wasn't the one who did the damage, but I was the instigator, so I guess I ruined it for the rest of the guys.

CHAPTER SEVEN

CANNED

In 1996 I fished opie season from January to March and then went cod fishing with my dad in April. Every once in a while, my dad would bring me up to the wheelhouse and show me how to haul the cod pots. He'd let me sit in the captain's chair and run the show for a bit. It wasn't much. Maybe ten pots a day, but it was valuable experience. It wasn't something I'd asked to do. He thought it was a good idea for me to start learning so I could keep moving up the ladder. He was always looking out for me. It took a long time for the light to go on, but I'm grateful that it finally did.

We didn't finish up the season until May. Out of the previous seven months I'd had five days off.

Lisa didn't like it, and neither did I. I was starting to get burned out. I was using my wages to pay off the previous year's taxes. Not only was I not making any progress on paying off the duplex, I wasn't putting any money aside to pay the taxman. I was caught in a vicious cycle.

I took the month of May off and finally got to spend some time with Lisa and Stormee, but that summer I went tendering again. It was a difficult time for my dad because he and my uncle were constantly at each other's throats. Normally one of them would run the boat for two months and then the other would take over for the next two months, but after a couple of lean seasons neither one of them could afford to take that kind of time off. Instead of rotating on and off, they ran the boat together. That was a big mistake. They were constantly bickering with each other, and for a while things were pretty tense on the *Arctic Lady*. Nonstop conflict and controversy and everyone—myself included—was sick of it.

When I was with my dad all he talked about was my uncle. He was still struggling with his farm, and recounting his arguments with my uncle was the only outlet he had. Same deal with my Uncle Kevin. He was fully recovered from his staph infection and making up for lost time. Every time I was alone with him he'd tell me how stubborn my dad was. They were always threatening to dissolve their partnership. It was like being around an old married couple who were too stubborn to admit they should have gone their separate ways a long time ago. There was definitely something in the air.

One day we were in port and I was working on deck when some dude came walking down the pier.

"Hey, is Scott Campbell Jr. on the boat?"

"That's me," I said. "What's up?"

"I've got this envelope for you."

I stopped what I was doing and went over to see what this guy had for me. As soon as he handed over the envelope he said, "You've been served," and hightailed it out of there.

I opened up the envelope. It was divorce papers.

Lisa had filed for divorce.

I called her and demanded to know what the hell was going on. She said that she'd had enough of my being gone all the time and wanted out. I talked her into waiting until I got home, and I caught a flight out of Kodiak the next day.

Back in Walla Walla we tried to work things out. There wasn't much to talk about. As far as she was concerned, I had a very simple choice to make: fishing or family. She made it sound like a no-brainer, but it was much more complex than that.

For me, fishing and family were intertwined. It was more than just my livelihood—it was in my blood, a family tradition. I'd been around it my whole life. I'd practically grown up on boats. There was no way I could ever separate family from fishing and fishing from family. I'd be turning my back on my dad and my uncle. Besides, between the mortgage and my taxes, I was under a ton of pressure to make some money, and I couldn't make any money unless I made things right with Lisa.

Things were looking pretty bleak until the Alaska Department of Fish and Game announced the opening of king crab season. King crab was big money. I knew there was a very good possibility that I could make all the money I needed to pay off my taxes, pay down the mortgage, and put some money in the bank. It was the answer to my prayers. But I needed to convince Lisa.

I knew Lisa didn't like my long absences. It was starting to wear on me, too. I needed to get home and spend some time with my family. I had a baby girl I'd barely seen, and we were still having money troubles. Lisa didn't see the point of my being gone all the time if it wasn't solving our financial difficulties. I figured if I could just get one big season, it would fix everything.

At least that's what I thought at the time.

Somehow I talked Lisa into letting me go out for one last trip out to the Bering Sea—no ifs, ands, or buts. This was my last last chance, and this time I really meant it.

I was hoping I'd get to fish with my dad one more time, but he was going through a breakup of his own. He decided to leave the *Arctic Lady* to my uncle and skipper the *Lady Alaska* instead. So I went out on the *Arctic Lady* with my uncle to go king crab fishing.

When I had first started out, my uncle was really tough on me—with my dad's blessing. As I got more experienced and proved myself as a fisherman, the Colonel eased up on me. But my uncle never did. As far as he was concerned, I was just another deckhand.

At the end of every season, when everything was put away until our next trip and we were getting ready to head out to the airport, my uncle would come down from the wheelhouse and say, "We could have done better." Then he'd launch into all the things the crew should have done to make it a more successful season.

It was such a deflating feeling to come back from a great season—better than just about everyone in the fleet—and be riding the high that comes with knowing that all the hard work and long hours had paid off, and then have to listen to that bullshit. All I ever wanted was for him to come down and say, "Good job." But he never did. He was never satisfied.

That last season was no different. We had a good season. Not great, but good enough to get me out of the hole I'd put myself in. It took me three long years, but I'd finally done it.

I didn't care if my uncle thought I did a good job or not. When he came down to give his speech, I had two words for him, two words Lisa had been waiting years for me to say.

"I quit."

I needed to find a job and fast. Lisa was working full-time as an office nurse at St. Mary's Medical Center in Walla Walla, but it wasn't enough. Our duplex on 13th Street cost us considerably more than five hundred dollars a month. I needed to make some money.

I had a couple of friends who said they could hook me up with jobs. One worked for a helicopter logging company, and the other worked as a driver for Pepsi. Helicopter logging sounded a hell of a lot more exciting than driving a truck, so I decided to try that. I called the owner and told him I was a fisherman. He was willing to give me a shot. All I had to do was go down to the shop at six the following morning and meet with the mechanic, and he'd drive me out to the site. I couldn't believe my good luck.

I went down to the shop, but no one was there. I waited for over an hour, but the mechanic never showed. I tried calling the owner, but I couldn't reach him either. They'd blown me off.

I called up my buddy at Pepsi, and he told me to come on down to the plant. There was just one condition: I had to be clean shaven. So I went home, shaved off my fisherman's beard, and was at the Pepsi plant interviewing for a job by eight o'clock in the morning. I told the manager that I was used to long hours and ready to work.

"When can you start?"

"How about right now?"

He hired me on the spot. He trained me that day, and I was ready to start the following morning. When I got home that night, the owner of the helicopter logging company called and apologized for the miscommunication. He wanted to know if I could start the next day. I told him I'd already found another job. That mix-up may have been a blessing in disguise. The guy who took my place at the helicopter logging company was nearly killed on the job. A log fell on him and broke his neck.

Driving a truck for Pepsi wasn't nearly as dangerous or as exciting as crab fishing on the Bering Sea. The crab seasons were so long and arduous, I never thought I'd miss them, but I did. I missed the excitement of being up on top of the stack in the middle of a storm. I missed that moment when the pot swings over and is packed full of crab. I missed the feeling of accomplishment that came when we returned to port for the last time after a kick-ass season. Instead I drove around the Walla Walla Valley in my truck, listening to country-and-western songs on the radio, delivering pop to supermarkets like Albertson's and Safeway. I had to wear a uniform. It was boring as hell.

On the weekends, I tried to inject a little excitement in my life by going snowmobiling. I bought a snowmobile and turned into a teenage daredevil all over again. I was always making these crazy jumps or trying to climb impossible hills. One time I got my machine stuck in the top of a tree. I was going for some big air and didn't quite clear the trees. I knocked down so many trees, my friends called me "Chainsaw."

In a way it made me miss fishing even more. It was fun to go tearing around in the snow, but it didn't compare to being out in the Bering Sea in a sixty-knot blow and taking forty-foot seas head on. Now *that's* an adrenaline rush. Plus, I had been getting paid for it.

To go from working around the clock on the Bering Sea to punching a clock for Pepsi was kind of a culture shock. The work wasn't really work. It was just driving around, dropping off deliveries, shooting the shit with people. Eight hours would go by in a blink.

I wasn't the type to go home at the end of the day and sit on the sofa all night, either. So I'd meet up with my friends at the bar and have a couple of cocktails. Most of the time I'd go home after that, but sometimes I'd get carried away and stay out all night partying. There were times I'd go back to work in the morning wearing the same clothes I'd worn the day before. My boss would pour coffee down me to get me to sober up. He liked me because I was the hardest worker they had, but I was also the biggest hell-raiser.

Lisa didn't care for it one bit. I thought that being home would ease the tension between us, but it made things worse. I was partying too much and wasn't being a dad. Our main problem was that we had no communication. I was very stubborn. She was very stubborn. It was a volatile combination. I was like a stick of dynamite. You could kick me around and toss me about, but once the fuse was lit, it was just a matter of time before I went off.

I quit fishing because I didn't want to lose Lisa, but being home all the time made me realize how much I missed fishing. After three years on the Bering Sea, I was still a hellion who, left to my own devices, made a lot of questionable decisions.

Like staying out all night.

Like drinking and driving.

It was only a matter of time before my bad decisions and reckless behavior would catch up with me.

Then it happened. I got pulled over one night and was charged with driving while intoxicated. That was the last straw. Lisa took Stormee and moved in with her mother. It was the first time we separated.

I was able to fight the DWI charge and get it knocked down to negligent driving, but I had to hire a lawyer and it wasn't cheap. Another costly mistake. After about a month apart, Lisa and Stormee came back.

I don't know if Pepsi found out about my arrest or if I was part of a random sweep, but I got called in for a drug test.

I panicked. I wasn't a habitual pot smoker, but sometimes when we were partying, a joint would get passed around and I wouldn't say no. I knew there was no way I was going to pass, so I came up with another solution. I had a buddy pee in a cup for me and pulled the old switcheroo.

They say it's only cheating if you get caught. Well, I got caught. Big time. The urine in the collection cup wasn't the right temperature, so it didn't work. I had to take the test again—this time with a pecker checker to make sure everything was on the up and up, so to speak. My boss had to be in the bathroom with me while I took a leak. Talk about awkward.

It turned out that the second test was unnecessary. Because I'd been caught cheating on a drug test, my Commercial Driver's License was suspended for a year, which meant I could no longer do the job I'd been hired to do. My boss fought to keep me. He put me in the plant stocking shelves, but there was nothing he could do. The head honchos at corporate wanted me gone. They had a pretty strict policy about "tampering with drug detection and collection materials," so my boss was forced to let me go.

I got canned by Pepsi.

To this day it's the only job I've ever been fired from.

The funniest thing about the whole fiasco was I got the results from the drug test in the mail a few days after they gave me my walking papers. I'd passed the test. I went through all that trouble for nothing.

It couldn't have happened at a worse time. Lisa and I were working things out—barely—but still heading in the right direction. She'd moved back into the house, and things weren't great but they were better than they'd been in a while. I'd cleaned up my act and was walking the straight and narrow while I got my legal issues sorted out. Getting fired from Pepsi was the last straw. Lisa moved out and started the divorce proceedings again. This time there was no talking her out of it, and I can't say I blame her. What kind of man can't hang on to a job as a soda pop deliveryman? What kind of man had I become?

I had no job, no wife, and no driver's license. I had nothing to do all day except sit around the house and feel sorry for myself. I figured it was time for me to go fishing again.

Right after I quit the boat, my dad had a monster season on the *Lady Alaska*. Each member of the crew made between eighty thousand and ninety thousand dollars apiece. It was one of his best seasons ever, and I missed it. I called him to see if I could get back in the action.

"If you're going to go fishing again, you need to get serious about it."

"I am serious," I protested. "I really want to do this." But that's not what he meant.

"You need to get your license and work your way up to being a captain."

Captain.

It felt strange saying that word and even stranger coming out of my dad's mouth. Was this the same guy who had tried to keep me from going fishing? And now he wanted to get me into the wheelhouse?

It meant my dad respected me as a fisherman, and the feeling was mutual. Now that I had some seasons under my belt and had observed how different skippers went about their business, I appreciated my father's knowledge, experience, and determination. He wasn't the easiest guy to work for—they didn't call him the Colonel for nothing—but when my dad was in the wheelhouse, you knew you were going to make some money. It didn't matter if it was the *New Venture,* the *Arctic Lady,* or the *Lady Alaska.* No matter what boat he was running, it had its best seasons when my dad was at the helm. So if he said I should get serious about becoming a captain, he wasn't just blowing sunshine up my ass because I was his son and had mouths to feed. If my dad said it, he meant it.

I packed my bags and went fishing with my dad on the *Lady Alaska*. During a three-week break between the seasons, I enrolled in a course in Kodiak. In order to get the necessary insurance to run a fishing vessel, I needed a nautical license.

Lisa and I were trying to work things out. I didn't want to take any chances that she might change her mind if I left her behind, so I took her to Kodiak with me. Just the two of us. Stormee stayed home, and my mom and mother-in-law split time spoiling her. It was Lisa's first real break from being a full-time mom. I didn't realize how much she needed to get away. While we were in Kodiak together we were able to spend some time trying to figure out what we could do to make things better between us. It was a great experience for both of us, and I kicked myself for not having thought of it sooner.

I passed the course and got my nautical license. The ironic thing about it was I couldn't drive a Pepsi truck, but I was qualified to captain a one-hundred-ton commercial fishing vessel.

Now I just had to find someone willing to take on a freshly licensed skipper. It was the age-old problem: I needed some experience before anyone would hire me, but to get that experience I needed someone to give me a chance. It was like being a newbie all over again.

Once I had my license, I started spending a lot more time in the wheelhouse with my dad. He didn't hold anything back this time. He put me in the captain's chair and taught me everything he knew. As salmon tendering season approached, my dad thought I was ready to handle the boat on my own. The owner of the *Lady Alaska*, however, needed convincing. He was a super straight-laced guy. Extremely cautious and conservative. He wouldn't let just anyone run his boat. You had to be a highliner, a top producer. When my dad told him I would be taking over for him that summer, he just about lost it. He was beside himself. I didn't blame him. License or no license, I was just a kid and he wasn't going for it.

But then fate intervened.

The *Lady Alaska* was scheduled to go down to Seattle later that summer to be converted for longlining brown crab. A boat that was equipped for both single pot fishing and longlining could participate in more fisheries and make more money. With the crab seasons getting leaner, a lot of boats were converting to longlining so that they'd have more opportunities to fish. With the *Lady Alaska* going into the yards, that meant the boat was going to be available for tendering for only about twenty days, which was a real short period of time. So short, in fact, that the owner wasn't able to find another captain to do it. Finally the owner relented and gave me a shot. That's how I got my first experience as the captain of a fishing vessel.

Tendering is easy work. I was in protected waters the whole time, but that didn't stop the owner from calling me every day. He was checking on me to make sure I wasn't scratching up his boat.

"Don't worry," I told him. "Everything is fine."

We were tendering red sockeye salmon. When my tanks were full, I headed for a cannery located on the Naknek River in Bristol Bay. It was a real shallow system and the tide was going out.

The prudent thing to do would have been to wait for the tide to come back in before heading up the river. You never want to go upriver on a falling tide, because if you run aground, you're screwed; you have to wait for the tide to come back in again before you can get loose. You always want to go in on a rising tide so if you do get stuck you can rise up with the tide. Well, prudence was never one of my strong suits.

I decided to be a hot shot and shoot up to the cannery before the tide went out. I got the boat stuck halfway up the river.

This was bad news. I had no idea how low the tide would go. I could end up sitting high and dry on a sandbar and damage the boat. Or, when the tide came back in, it could roll the boat over on the current and then I'd really be in trouble. Plus, I had a full load of fish I needed to get to the cannery. Last, and certainly not least, it was embarrassing as hell to be stuck in the middle of the Naknek River for everyone to see.

I had to act fast. I thought if I blew all the water out of my tanks, that might get me over the bar. That water, however, was cooling the salmon. If it didn't work, the fish would spoil and I'd have a boatload of rotten fish that I wouldn't be able to sell. I didn't even want to think about that possibility, so I stopped thinking about it and blew the tanks.

Talk about an agonizing few minutes. As the water drained from tanks, the tide continued to fall. I thought about all the times I'd put myself in this position. I'd wrecked motorcycles, all-terrain vehicles, pickup trucks, and snowmobiles. If it had a motor and gears, I'd destroyed it. Now that "oh shit" feeling was climbing up the walls of my gut as my body tried to catch up with what my brain already knew: I'd fucked up again. I remember the feeling from when I was kid, sitting upside down on my ATV wondering what the hell had happened. I had it the night I got drunk with the boys and tried to run my truck up a hill that was too steep and slippery. I thought I'd make it then, too, right up until the moment when my momentum stopped and my truck was perfectly still, like a killer whale doing a trick at the aquarium before tumbling down the hill and rolling on its side. That helpless feeling came over me, and it felt exactly like being a teenager in trouble. I'd done it *again,* and I had no one to blame but myself. When was I going to learn? When was I going to stop being my own worst enemy and stop taking such foolish risks?

All I could do was hope and pray that the water would get out of the boat faster than it left the river.

It worked.

The boat rose up in the water just enough to get me over the bar.

Now I had a new set of problems. I had no water on my fish. When I got to the cannery, I had to put "hot" water back in the tanks—water that was considerably warmer than the water that had been in the temperature-regulated tanks.

Bottom line: I'd compromised the quality of the fish.

I got my ass chewed on that one. I was trying to be a hero and ended up costing the boat some money.

I learned a lot of valuable lessons that summer. Unfortunately I had to learn them the Campbell way.

﹋

When my twenty days were up, my dad came back to the boat and we took the *Lady Alaska* down to the shipyards in Seattle, where a guy named Bill Widing took over.

Bill was an interesting guy. He had experience with brown crab, but he was diversified and owned his own boat, the *Amatuli*. The owner of the *Lady Alaska* offered the brown crab fishery to my dad first, but he'd never fished brown crab before and didn't feel comfortable doing it, so he turned it down. When the owner suggested that he work as the relief captain, my dad packed up his stuff and left the boat. He was the captain or nothing. The Colonel had too much pride to play second fiddle to anyone.

The first thing Bill did was fire everybody on the boat—the entire crew—except me. I was the engineer; he needed me. He didn't want a bunch of guys who didn't have experience in the fishery, so he brought in a crew who knew how to fish brown crab and I got a crash course in longlining. I was used to single-pot fishing, where each pot has its own line, but I'd been longlining on the *Beauty Bay*. With longlining all the pots snap onto a single line. It was a totally different style of fishing, and I didn't know Bill all that well, but I was up for it.

We had a pretty good season—so good that Bill made me an offer when it was over.

"If you come over to the *Amatuli* and chief my boat for opie season, I'll let you run the boat yourself for cod season."

I couldn't believe what I was hearing. This was it. This was my big break, a chance to prove myself in a fishery, establish a name for myself. It wasn't a crab fishery, but I didn't care.

I called my dad to tell him the good news. To say I didn't get the reaction I was looking for would be an understatement.

"You're not ready to run a boat by yourself," he said.

"What do you mean? Of course I'm ready."

"Fishing and tendering are totally different animals."

"You think I don't know that?"

"You'll know when you're ready and you're not ready."

I told him I didn't see it that way, but deep down I knew he was right. There was so much I still didn't know, so many ways I had yet to be tested. But ready or not, I knew that opportunities like this didn't come along very often. What was I going to say, "Thanks, but no thanks. I'm not ready for the responsibility"? I was Bill's relief skipper. If I told him I couldn't hack it, he would fire me on the spot. As his back-up he needed to have total trust and confidence in me. As far as I was concerned, I couldn't say no. If I wanted to take the next step and be taken seriously as a skipper, I had to go.

My dad came back to the *Lady Alaska* for opilio crab season, and Bill and I went out on the *Amatuli*.

It was a strange feeling working on a boat without a Campbell in the wheelhouse. Even though I'd worked with Bill on the *Lady Alaska*, I still thought of it as my dad's boat. It was my first season away from the family operation.

The *Amatuli* was a step down from the *Lady Alaska* and the *Arctic Lady*. Make that several steps down. It was an older boat and looked like it hadn't been in the shipyards in a long time. When my dad was the skipper of a boat, he looked after that boat as if it were his own. I thought this was standard operating procedure in the fleet, but that turned out not to be the case. Most captains cared about the fishery first, and the boat came in a distant second. The strange thing about the *Amatuli* was the skipper was also the owner, and he'd let his own boat fall into disrepair.

Over the course of that first opie season on the *Amatuli* Bill ran me ragged. I went from the deck to the engine room to the wheelhouse. It was a never-ending cycle. I spent just about all my downtime in the engine room trying to hold that junker together. As soon as I'd get one thing running right, a problem would spring up somewhere else. It was a smaller boat, and it wasn't the easiest to work on. I hardly got any sleep that season, but we made a little money. The main thing was, Bill saw enough in me to trust me as a skipper and turned the boat over to me for cod season.

Cod wasn't a big moneymaker in those days. It wasn't considered a key fishery like it is today. It was regarded as a filler season, a way to keep the boat busy and make a little money in between crab seasons. A lot of crabbers wouldn't go cod fishing. You had to work a lot harder for a lot less money. I knew firsthand how hard it could it be after I had gone cod fishing on the *Beauty Bay* and made a whopping fifteen dollars a day for three months. Because of that, a lot of skippers used a second crew for cod season—a B team—guys who weren't as experienced but were willing to work in the hopes that there would be an opening with a crab crew down the road.

Crab was king. That was the fishery that could make or break a boat. If a boat went backward during crab season, i.e., if it lost money, it could bankrupt an owner. No owner would trust a greenhorn captain to run the boat during crab season. There was just too much at stake.

If crabbing was the major leagues, cod fishing was the minors. It was the perfect scenario for a guy like me to come in and figure things out. If I failed, it wasn't the end of the world, because the paydays weren't all that big compared to crab. But I wasn't planning on failing. I was going to show the world that I had what it took to be a successful skipper.

I got off to a rough start. I couldn't convince any of my friends or colleagues to come fishing with me. During crab season there would be guys lined up on the docks waiting for a spot on a boat, but cod season was a different story. The fishermen I'd worked with before wouldn't sign on with me because I was a greenhorn skipper. The guys who worked the boats weren't that much different from the guys who owned them. They wanted to fish with a proven producer. They didn't want to spend a season busting their hump and not get paid at the end.

It was understandable, but it was also frustrating and humbling. Nobody wanted to take a chance on me, which made me want to prove them all wrong. It wasn't that they didn't think I could do it, but they wanted to see results. Fishing has always been a bottom-line business.

I got desperate and hired anyone I could. I had to hire processors, cannery workers, guys who had little or no experience on fishing vessels. I had to settle and go fishing with an inexperienced crew because I couldn't get any reputable fishermen. We were a sorry-looking lot, as ragtag a group as I've ever seen on a fishing boat. A greenhorn crew for a greenhorn skipper and a boat whose best days were in the past.

It was a difficult season. I didn't make just a few mistakes, I made all of them. Each mishap or error was compounded by the fact that no one on deck knew what he was doing. On top of that the boat kept breaking down, and because I didn't have an experienced engineer on board, it was on me to make the repairs.

We caught some cod, but our loads were always light. And every time we came into town, I'd lose another crewmember. I'd waste two or three days in town looking for replacements. As a last resort I'd prowl the bars looking for fishermen who were desperate enough to come fishing with us.

It was a nightmare season.

Our grand total was four hundred thousand pounds of cod. There were fishermen who caught over a million pounds that season. I didn't catch even half of what the highliners caught, but somehow the boat broke even. That was good enough for Bill, but it wasn't good enough for me. I knew I could do better. If I wanted any kind of career, I *had* to do better.

Incredibly, Bill was willing to give me another shot—this time during crab season. He knew his boat was in bad shape, and I think he was impressed with the way I'd managed to hold it together with a greenhorn crew. I knew I had what it took to succeed. My confidence hadn't wavered in the least, but I couldn't risk another season with a run-down boat and an inexperienced team. So I put all my energy into getting some quality guys to go fishing with me. I wasn't going to make the same mistake again.

Instead of focusing on the crew, I should have paid more attention to the *Amatuli*, because it was the boat that ended up biting my ass.

Make that finger.

CHAPTER EIGHT

DOWN TO THE BONE

Seven days.

That's how many days I had to salvage my reputation as a skipper, but the doctor dressing my mangled finger at the clinic in Dutch Harbor didn't care.

"You've got an exposed bone. You don't want to risk getting an infection."

The way I saw it, I could go home and sit on my sofa for seven days, or I could go fishing, get my quota, and save our season.

Seven days, that's it.

After the accident my crew thought I was crazy when I told them I wanted to get the gear out of the water before heading back to Dutch Harbor to get my finger looked at. But when the compressor chomped off my finger, we were just a few hours from the gear. Finger or no finger, the ice was coming down, and if we didn't get the pots, we'd lose them. One hundred twenty-five pots at two thousand a pop is two hundred and fifty thousand dollars. There was no way I was leaving a quarter of a million bucks in the water. As long as I was able to block out the pain, I was going to get those pots out of the water and bring them back to Dutch Harbor.

Fishing is a lot like football: Getting hurt is part of the game. Of course there's a difference between getting hurt and getting injured. When you're injured you're injured. That's when it's time to come out of the game. There's no shame in sitting on the sidelines when you're injured.

But everybody has to play hurt. Whether it's a blindside hit or a wall of green water, a 250-pound lineman or a 700-pound pot, if you play the game the way it's meant to be played, you're going to get knocked around. You don't play football or go fishing because it feels good. You play to win. By the same token, you don't quit just because it hurts. You fight through the pain and find a way to keep going. There's no heading for the sidelines when you're in the middle of the Bering Sea.

That's how I felt about my finger. I was playing a game I couldn't afford to lose. My entire career was on the line. As far as my finger was concerned, I was hurt, not injured. I had no intention of coming out of the game. I'd promised the crew I would do everything in my power to make the season a success. Calling it quits on account of a busted finger just wasn't going to cut it.

After the accident I downplayed the situation to avoid freaking out the crew. But I'm not going to lie: The pain was intense. I knew that I had to keep my hand elevated above the level of my heart so that it would stop bleeding. The tip of my finger where the bone was showing throbbed with pain. I thought of soldiers on the field of battle who'd sustained injuries far more devastating than this and still managed to carry on and continue their mission. I wrapped my finger in some bandages in the first aid kit and went full steam ahead for the fishing grounds. We got up there, moved our gear, and within a matter of hours we were headed back to Dutch Harbor.

While we were out, the strike that had delayed the season had been settled, and the fleet was on the move. We passed them as we came into port. I thought it might speed things along if I radioed ahead to let them know I was coming in and that I needed medical attention, but I knew that my dad would be listening in and I didn't want him to worry about me. He had his own job to do.

Seeing the fleet heading out to the fishing grounds got my competitive juices flowing. I did the math and figured the season would last about seven days. Seven days was nothing. I could do seven days on one leg. How hard could it be minus part of a finger?

At the clinic the doctors predictably didn't want me to go back out. They were astounded that I had any desire to do so.

"An exposed bone is a bad deal! You need to get on a plane and get to a hospital!"

They tried but they couldn't convince me. I knew that if I could just hang in there for seven days, I could save my season.

"What are they going to do for me in Anchorage?"

"Give you antibiotics."

"Do you have the antibiotics here?"

"Well, yeah . . ."

"Then give me the antibiotics."

I loaded up on drugs and dressings and went back to the *Amatuli*. Because of the strike the crew had been on the boat for a month, and we hadn't made a nickel. With the fleet on their way to the fishing grounds, there was no way my guys would find another captain for the boat or a spot on another vessel. If I didn't go fishing, I'd screw the crew, the owner, and myself, because there was no way someone would give me another shot at running a boat.

So against doctor's orders I decided to finish out the season. I called Lisa and did what I always did: I lied. I told her the doctors said it wasn't that big of a deal and I just needed to keep it clean until it was time to come home. Lisa didn't care what the doctor said. Finger or no finger, she wanted me home.

As a nurse she knew a hell of a lot more about this stuff than I did, so I made it sound really minor, which to my way of thinking, it was. If the accident had happened a few years later, she would have demanded that I text her photos of my finger. It's a good thing we were a couple of years away from that technology, because in truth my finger looked pretty gnarly. She would have taken one look at the bone poking out of my finger and booked my flight home on the spot.

Even though it hurt like hell, I didn't see what the big fuss was all about. It wasn't like I'd lost an arm or a leg. It was just a finger, and not even the whole finger but part of it. I'd be back in seven days and that would be the end of it.

What could possibly go wrong?

———

Pain is like the sea. There are ups and there are downs, and as soon as you think you've got things under control, it knocks you flat on your ass.

That's how the pain in my finger was. There seemed to be no rhyme or reason as to why it would flare up. I'd be sitting in my chair, steaming along, and out of nowhere the pain would shoot through my arm. I wasn't taking anything for it, because I couldn't be all zoned out while I was driving the boat. I had no choice but to bite down and bear it.

The worst pain of all was when I whacked it on something down in the engine room. The pain was so intense I would actually see stars. One good thing about being in the engine room was that the noise down there was so loud that no one could hear me scream. I was still serving as my own engineer, so I had to go down there when the boat broke down, which it did.

Every. Single. Day.

That slowed us down considerably. Seven days turned into fourteen.

If it wasn't the boat, it was the weather. I quickly discovered that the *Amatuli* wasn't made to handle extreme weather conditions. We were fishing in an area with a lot of current. Bad weather plus a strong current meant massive seas. I thought I could fish through it.

I was wrong.

We took a giant wave on the forward part of the boat. It just rose up out of nowhere and slammed into a deckhand named Steve Wilson. The wave took him at the bow, knocked him across the deck, and slammed him into the house—a distance of about sixty feet. He was washing around on the deck like a rag doll.

I jumped on the loud hailer.

"Is he dead?"

The crew rushed to his side. "He's alive!" they called up.

I don't know how he survived, but that wave banged him up pretty good. He busted some ribs but still fished the rest of the season. Probably one of the worst injuries I've seen someone come back from—aside from my own. Tough kid. After that season he never came back.

The ice that came down out of the Arctic Circle and covered the fishing grounds forced us to try our luck elsewhere, and the fishing was horrible. No matter low long we let the pots soak, we were catching hardly anything. It wasn't just me. The whole fleet was struggling. I listened to the radio for clues as to where the crab were running, but either the captains were more secretive than usual or no one was having any luck.

Fourteen days turned into twenty-one.

At some point during the third week, my hand began to swell. I spilled some diesel fuel in the engine room, and it ran down my arm, soaked my dressings, and got into my wound. Not good. That did a number on me, and soon afterward my hand was tender, puffy, and sore. It looked like a pimple that was waiting to pop. My hand was starting to look like my uncle's knee, so I stopped looking at it, even when it swelled up to twice its normal size. It got so bad that I couldn't use it anymore. I had to tape my hand to the throttle like a demented sea captain who lashes himself to the wheel during a storm. I got so weak I couldn't get out of my chair. For the last four or five days of the trip I never left the wheelhouse. The guys would bring me meals, and the only time I got up was to go to the bathroom. I was in a terrible state, but I refused to quit.

Finally, after twenty-eight days, they announced the closing of the season, and I collapsed. I couldn't take it anymore. What was supposed to be a week of fishing had turned into a short month. The guys carried me into my stateroom and drove the *Amatuli* back to Dutch Harbor. They got the boat tied up and put me on a flight home. Lisa was waiting for me at the airport, and she took me right to the hospital where she worked. She'd lined up appointments with the general surgeon and orthopedic surgeon. The general surgeon eyeballed the bone sticking out of my finger and sent me to the orthopedic surgeon. Less than an hour later, I was in surgery. I don't remember any of it. All I know is that I finished the season.

⸻

"You're not taking the hand," I said.

The doctor explained that I had a really bad staph infection in my finger. Even with surgery they couldn't get it out, couldn't stop it from spreading. Now they wanted to go in again. Only there was a 50/50 chance I was going to lose my hand.

"No way," I said. "I gotta have it. You can take the finger, but you can't have the hand."

I'd been in the hospital a week, and they weren't going to let me out anytime soon. At least this time I had health insurance.

"I'm not saying we will," the doctor said, "but we might have to. If we don't, you could die."

"Do whatever you gotta, do," I said, "but don't take my hand."

It was ridiculous. All because of a faulty air compressor motor I was going to die or lose my hand. Of course my own stubbornness played a part in my predicament, but it was easier to blame the boat.

The surgery involved a procedure called backflushing, where they'd insert an IV in my finger and flush the infection out. Basically they were going to blow out the tip of my finger and let it heal back up again.

All blood wants to go to the heart. To make sure the infection didn't travel back up into my bloodstream, they had to cut off the circulation in my hand by folding it down so that my fingers were touching my forearm. That's where it got dicey. If my hand didn't respond well to having the circulation cut off, they'd have to amputate it. I didn't really care how they did it, as long as they didn't take my hand.

I wasn't thinking clearly, and I let them put me under. I just wanted to get the operation over with and the infection out of my body as quickly as possible. It never occurred to me that I wouldn't be able to stop the doctors from cutting off my hand if I was unconscious.

I woke up groggy and dehydrated. It took me a while to process the bandages wrapped around my hand. A hand I could no longer feel. That's when I started to put the pieces together. I was certain the reason why I couldn't feel my hand was because it wasn't there anymore.

I started screaming and shouting and tearing off my bandages. "My hand! My hand!"

Lisa screamed back. "Stop it! What are you doing?"

"They took my hand!"

"No they didn't!" she insisted.

"Bullshit! They took it." I waved my arm around. All that was left was a bandaged club.

"It's there!" she said.

"You're lying!" I shouted. "I can't feel it!"

"Don't touch it!"

It was one of the stranger arguments we've ever had.

I was freaking out big time and wouldn't listen to a word Lisa said until the doctors came. They opened up the bandages a bit so that I could see my hand. I'd never been so relieved to see my fingers.

Just not all of them.

The doctor told me that they'd had to take off the tip of my infected finger. They clipped it off at the knuckle. After thinking I'd lost my whole hand, a knuckle didn't seem like such a big deal.

They kept me in the hospital for another two days, and then two months later I had to go back for more surgery—all part of the sacrifice I had to make to save the season, and rescue my career.

And it paid off.

While I was recovering I got a call from Fred Wahl, the owner of one of the newest boats in the fleet, called the *Vixen*.

He wasn't happy with the guy who was running his boat. He hadn't made any money with it, and he'd gotten four or five guys hurt. The skipper was running the boat into the ground, and Fred wanted to make a change.

"I want you to come run my boat," Fred said. "If you've got that kind of determination, I know you're not going to fail."

I was still recuperating from my injuries, and here I was being offered a bigger boat and a better opportunity. It was too good to be true. I accepted on the spot.

"There's just one thing," Fred added.

"What's that?" I asked.

"Don't get anybody hurt."

There was no mistaking the sound of the *Vixen*'s alarm. A loud piercing wail like a fire engine blasted throughout the decks. It had to be loud enough to wake up the crew down below or be heard over the winds of a raging storm. Mostly it startled the shit out of me.

It was a general alarm, which meant the crew had no idea what was wrong. As the captain it was up to me to consult the instrument panel in the wheelhouse and figure out where the alarm had sounded and what to do about it. But I knew what the problem was before I even got out of my chair.

The alarm in the bilges had been giving us fits all morning. There was a short somewhere along the line, causing it to go off. Once got your attention. Twice tweaked your nerves. Anything more than that meant someone wasn't doing his job.

I sent my engineer down to flush the bilges and fix the problem. I'd personally handpicked him for my second stint as a captain. After my adventures with the compressor on the *Amatuli*, I was done pulling double duty as captain and chief. I wasn't about to head out to sea again without a first-rate mechanic on board. I'd worked with this guy on plenty of boats, and he'd always struck me as a capable mechanic who knew his way around an engine room.

Not that I was expecting any trouble. The *Vixen* was one of the newest boats in the fleet. It was the antithesis of the *Amatuli*. It was big and shiny and had the latest gear that money could buy. The owner's expectations were sky-high, and I was determined to reward his confidence in me with the best opie season the boat had ever seen, but I was starting to have doubts about my chief.

His solution for fixing the short in the alarm was to disconnect the wiring from the bilges. That's like taking the batteries out of a smoke detector because it goes off when too much smoke comes out of the kitchen. Of course the engineer didn't tell me that's what he'd done. He was counting on the fact that I'd only find out if something went catastrophically wrong down in the bilges, which was exactly what happened.

It was our first day of our second trip of the shortest opilio crab season that anyone on the boat had ever been on. If the first three days were any indication, we were in the midst of an epic season. Every time we hauled our gear, the pots were stuffed with crab. Huge numbers. We kept waiting for the streak to end, but the crab kept on crawling into the pots. Sometimes they'd ride along on top as we brought them in. We filled the boat in three days. To fill a boat the size of the *Vixen* in such a short period of time was impressive. When I radioed the news to the owner, he was overjoyed.

We were headed back out to the crab grounds when I noticed that the bow was feeling a little sluggish. It wasn't very responsive and felt like it was riding too low for the kind of seas we were in.

I sent the engineer forward to investigate, and he came back with some bad news.

"The bow's flooded."

"What the hell happened?" I demanded.

He stared down at the deck. "Something must have happened when I pumped the bilges."

It was obvious what that something was: He'd left the valve in the sea chest open.

At the bottom of the bilges there's a valve called the sea chest. It probably sounds counterintuitive to have holes in the bottom of a boat, but they're necessary so you can get the water out. It doesn't take an expert to tell you that leaving a valve in the bottom of the boat open is a great way to sink your boat. Once you flush the water out, you shut the valve so that seawater doesn't get back in.

The engineer hadn't done that. For whatever reason he'd left the sea chest open.

Luckily the pumps hadn't been submerged, and we fired them up and got the water out pretty quickly. To speed up the process, we also engaged the emergency pump in the engine room. It took a while, but after a few hours we were back in business.

We were turning and burning toward our destination: the opilio grounds north of the Pribolof Islands, two big black piles of rock in the dead center of the Bering Sea. As we dieseled past St. Paul and St. George, I brooded over the bilges. It bothered me that the chief had left the valve open, and it bothered me that he'd disconnected the bilge alarm. You don't have to like the people you work with on a boat, but you need to be able to trust them. I was having some serious trust issues with my engineer.

I sent the engineer down to the bilges again to make sure everything was okay. A few minutes later one of the deckhands burst into the wheelhouse.

"The engine room is flooded!"

The deckhand had come in from topside and saw the engineer standing in the doorway to the engine room, watching the water rising in the compartment below.

I knew in an instant what had happened: When the engineer had fired off the emergency pump to clear the bilges in the bow, he'd left the valve open in the sea chest.

Again.

It was the only explanation. The *Vixen* was a new boat. She hadn't ruptured a seam or had a mechanical failure. And the engineer had done the very same thing three hours earlier.

Only this time the consequences were much more dire. The flood wasn't in the bow, but in the engine room where the two main diesel engines and generators were located. If we lost power to the mains, we were done for. Without power there'd be no way to pump the boat. We'd just keep taking on water until we couldn't take any more and the brand-spanking-new *Vixen* would sink to the bottom of the Bering Sea.

The first thing I did was what I always did when I was in big trouble: I called my dad.

"We're taking on water in the engine room," I said.

"How bad is it?"

"I don't know. I'm about to find out."

I gave him our position so he'd know where to find us if we sank.

"Be careful," was the last thing he said.

I jumped out of my seat and ran down the ladder to the engine room. The engineer was still standing there. Even though two or three minutes had gone by since the deckhand had told me the news, the engineer was completely immobilized. He just stood there in the doorway, looking at the water down in the engine room rising higher and higher as if he were mesmerized, like he didn't know what to do.

It was a bad, bad scene. There was six feet of water, and it was rising fast, already halfway up the main engines. Black smoke poured out of the diesels and the generators were sparking. I could feel the chill coming off the water. It was January, the dead of winter, and the water was thirty-four degrees Fahrenheit. It had only been a few months since my accident on the *Amatuli*. I'd done everything I could to avoid getting myself into another situation like this, yet there I was standing on the edge of disaster. And if I didn't get that valve at the bottom of the bilges shut, the boat would sink and we'd all go down with it.

I dove in.

What does thirty-four degrees feel like?

A complete and total shock to the system. My instinct was to yell, but if I did that I'd drown. I had to stay calm. I had to focus. These things are required during any difficult task, but plunging into freezing-cold water was like a punch in the face. Thirty-four degrees Fahrenheit is just two degrees above the temperature that freshwater freezes. I could feel my body go numb as my brain teetered on the edge of panic.

I could see surprisingly well. The overhead lights shone through to the deck, so I could see where I was going. I wasn't worried about drowning. I didn't think about getting electrocuted. But if the lights went out while I was underwater I would be screwed. Everything would turn pitch-black. It would be like being inside an underwater cave. Finding the valve would be impossible. I wouldn't know which way was up or down. It would be easy to get disoriented, panic and lose my shit.

I didn't want that to happen. Lost, scared, cold, and confused is a terrible way to die. I went straight to the bilges. Because the boat was new, the valve turned easily and I was able to get it closed. I swam to the door and climbed out of the engine room. My crew was already pumping the water out. We'd saved the *Vixen* from sinking, but all I could think about was going to my stateroom and getting out of my wet clothes.

I grabbed some coffee and went back to the wheelhouse. My hands were turning purple and my legs were stiff and numb. I'd only been under a few seconds, but it had done a number on me. I paced the wheelhouse, trying to shock some life back into my legs.

I called my dad and told him that things were under control. He continued his course to my location. There was no way he was going to turn back. He'd notified his brother of my situation. My uncle had been fishing ten miles south of my position and he, too, was coming. I felt a sudden surge of gratitude that my dad and my uncle were on their way.

Which was more than I could say for my engineer.

He'd made three serious screwups in a single day, each mistake more costly than the last. And when our ass was on the line, he'd frozen up. Something had to change.

I called the deckhand who'd alerted me about the situation in the engine room up to the wheelhouse and told him I needed someone I could trust to keep the boat running. The funny thing about this guy was that he was the last man hired. He was a career deckhand who loved fishing and had no interest in climbing the ladder. He'd hop from boat to boat, always looking for the best situation, which concerned me. I didn't quite know what to think of him when I hired him, but I was glad I did.

"I'm relieving the chief of his duties. From now on the job's yours."

He nodded. The crewmember who was almost an afterthought was now my right-hand man. He took no pleasure in taking another man's job, but he understood that you only get so many chances on the Bering Sea.

My father and uncle arrived on the scene and stood by while we got the last of the water out of the bilges. This time we remembered to close the sea chest.

But we weren't out of harm's way yet. Far from it. We had gear in the water that we needed to get on board. That's probably what saved our ass. If the gear had been on the boat, several tons of crab pots would have added to our tonnage. Those extra pounds would have caused us to take on more water and sink faster.

However, what had been a blessing was now a liability. I didn't want to risk taking on water again while we hauled the gear, but I didn't want to leave it behind either. I called the owner and told him the situation. He gave us his blessing to leave the gear, come back to port, and make sure the boat was in good working order.

I thought about it. If we went back now, we'd miss out on some of the most incredible fishing I'd ever seen. I decided to stick it out. I waved good-bye to my dad and uncle and got back to work.

The first thing I did was find the short in the bilge alarm. It took me three hours to trace the break, but I finally found it and got it fixed. After that I felt a lot better about our situation. At least if the bilges failed again, we'd have some warning.

We went back to what we'd come to do: catch crab. We hauled the gear on board, and each pot was stuffed with opies. Every string we hauled had the same result. Day after day of epic fishing. We filled the boat a second time.

I learned a few valuable lessons during that trip. At the beginning I thought my engineer was reliable and the deckhand untrustworthy. But when shit got serious, the opposite turned out to be true.

It was like the adage about the Marine on the firing line who knows all the tactical maneuvers backward and forward, but locks up when the bullets go whizzing past his ear. You just never know how a crewmember is going to act until you get him in a crisis. Then you find out what he's made of—if it's not already too late.

In the end the only people I could trust, I mean truly count on, were my dad and my uncle. Even in the middle of the Bering Sea, they were there for me when I needed them most.

The boat made more than a million dollars that season. This was at a time when half that much was considered pretty good. The owner was so impressed that he offered to make me a partner on his boat. I told him I wasn't in a position to make that kind of financial commitment, but I would be in a few more seasons, and we left it at that.

I felt like my career was starting to take off. I wasn't quite sure where I wanted to be, but I'd gone from bait boy to a potential partner in a few years' time, and that was something. It took risking my health and sacrificing my finger, but it seemed to be paying off.

By turning a potentially disastrous season for the *Amatuli* into a success, I was able to establish a reputation as a skipper who refused to quit. That was my attitude going forward. No matter what, I wasn't going to fail. I would not accept defeat, and everyone on board, from the bait boy to the engineer, knew it. No matter now bad the situation was, I'd find a way to succeed.

In a few short years I'd reached the pinnacle of my profession: I was a highliner, the skipper of a top producer. I'd officially made it. I'd paid a steep price, but it was worth it.

Now it was time to enjoy the fruits of that success. Little did I know that I'd pay a price for that, too.

CHAPTER NINE

THE *SEABROOKE*

While I was kicking ass on the *Vixen*, my dad was looking for a new opportunity. On the *Arctic Lady*, he'd been a part owner of the boat, but on the *Lady Alaska* he was strictly a skipper. He was looking to buy into a boat, and while scoping out the situation, he didn't like what he saw.

The fisheries were getting smaller and smaller. There was too much competition. Not only was it harder to make a living out there, it was taking a toll on the resources. He'd seen the fisheries fail in Kodiak—that was what brought him out to the Bering Sea in the first place—and he didn't want to go through that again, but he didn't think the state would let it happen.

To my dad's way of thinking, that could only mean one thing: regulation. Sooner or later either the federal government or the Alaska Department of Fish and Game would step in.

That wasn't necessarily a bad thing—as long as he was ready—and he was anything but.

My dad wasn't prepared to retire, but he was thinking about it. He was thinking about a lot of things. He was thinking about his family. He was thinking about his farm. He was thinking about me.

He knew that I was itching to move up the ladder, but he didn't believe I was ready to be a full-time skipper. When I told him my plans to buy into the *Vixen*, he suggested an alternate course.

"Let's partner up. I've still got some years left in me, and you need some grooming before you're ready to cut loose on your own."

"I don't know," I said. "I really like the *Vixen* . . ."

If he'd asked me a season earlier, I would have jumped at the opportunity. But I thought I had a good thing going with the *Vixen*. The *Vixen* was new, a quality boat. I'd made money on the *Vixen*. More important, my success with the *Vixen* didn't have anything to do with my dad or my uncle. It was my own deal. I thought the best way to keep building my reputation and establishing a name for myself in the fisheries was to keep doing what I was doing on the *Vixen*.

But my dad wasn't taking no for an answer.

"Let's go look at some boats," he said.

My dad could see the writing on the wall. There was a lot of talk among the fishermen about new policies and programs that would change the way the fisheries were regulated. It was only a matter of time before politics came into play.

The crab fisheries operated under an open-access policy. It was a system that encouraged all kinds of people to participate in the fishery. From outlaws to overachievers, party boats to people who only cared about their paychecks, there was a Gold Rush feel to the fishery. Derby-style fisheries reward risk takers and rebels. It was a broken system.

For one thing, the fisheries were overcrowded. In its heyday there were approximately 375 boats, and each boat was putting about 200 pots in the water per day. That's 75,000 pots being hauled every single day of the season. That was an awful lot of gear to drop on the crab. Because the fishermen were competing with each other, there was no time to let the gear soak. Each fisherman was trying to get as much crab on his boat as possible, because once they reached the quota, Fish and Game would shut the season down.

With that many boats in the water, a fishery might only last a couple of weeks. One only lasted twelve days. We had to get in, get what we could get, and get out.

Naturally there were always some fishermen who would ignore the guidelines and break the rules. Some skippers pushed their boats to the limit, and then they pushed them some more. They'd overload the vessels with too many pots. Guys would get hurt. Boats would go down. It was an unsafe fishery.

It was also extremely difficult to manage. Fish and Game relied on reports from the captains to determine when to cut off the fishermen. Some of the less scrupulous skippers in the fleet would lie about their numbers or report late to buy extra time. If the fleet went over the quota, and it almost always did, the Feds would come down hard on Fish and Game. This created an adversarial relationship among all parties in the fishery.

Plus some of these guys didn't give a damn about the resources. They didn't protect the small crabs and female crabs—the kinds of crab we were prohibited from fishing. The canneries couldn't take them, so these fishermen would smash the smalls and kill the females so they wouldn't end up in their pots again. It was a barbaric style of fishing.

There was a sizable percentage of the fleet who risked their lives and damaged the crab population for paydays that weren't very big and were getting smaller every year. No one was making the big money like they did in the old days. There were simply too many boats and too few crab. If things didn't change, it would end badly for everyone.

My dad was positive that Fish and Game was going to implement a rationalized fishery. That meant quotas would be established for boats that had been in operation for a sustained period of time and could demonstrate a record of responsible fishing. In rationalized fisheries each boat was given a quota that reflected their fishing history, like a handicap in golf.

If this system was implemented in the crab fishery, my dad thought it would do two things: cut way back on the number of boats and eliminate competition among the crews. This would result in a safer fleet and a more manageable fishery. It was good for the crab and good for the fishermen—as long as your quota was big enough to support your operation.

Programs that had gone into effect in other fisheries rewarded boats with a proven track record. My dad believed that established boats would have a leg up on boats that were new to the industry. The key word was *boats* not *fishermen*. The quotas would apply to boats and not the skippers who worked them. So if we didn't buy a boat before the new rules went into effect, the prices would go up and we would never be able to afford it. For instance, a commercial fishing vessel with a proven track record valued at a million dollars could be worth ten times that amount once the program was structured. Simple supply and demand. My dad was afraid that if he waited too long the market would dry up. Owners would hang onto their boats until the open-access era ended and the next chapter began.

If we wanted to get in on the new program, we needed to get a boat and we needed to do it fast.

Needless to say, my dad was in a bit of panic. He knew what was coming. He knew he needed to get a boat, but there weren't many available, and he had a very limited budget.

That's where I came in. He called and put the screws to me.

"Why don't we join forces and put our resources together? Instead of getting everyone else rich why not get a boat and stake our own claim?"

Why not? Because my dad wasn't the easiest person to get along with.

I remember when I was a kid my dad would get into fights with my grandpa. They argued about the farm, and I had to listen to it on both ends.

"Your grandpa this . . ."

"Your dad that . . ."

I'd be working on the farm, and my dad would tell me to do something one way. Then my grandpa would come along and tell me to do it a different way. Sure enough, when my dad came back and saw me doing it the "wrong" way, he'd give me hell for it.

They'd get into horrible arguments all the time and wouldn't talk for weeks afterward. After a while my grandpa would come down to our house and apologize, and things would be good for a few weeks until the next dustup.

The same thing happened on the boat when my dad was bickering with his brother. They'd disagree about something, and I'd get caught in the middle. They wouldn't fight in front of the crew. But when I was alone with my dad or my uncle, they'd complain about each other. In the end all that acrimony took a toll and they dissolved their partnership. That was the reason why my dad wanted to team up with me: He couldn't get along with his brother anymore.

Different generations, but the same stubborn Campbell streak. And I had it, too.

I never doubted for a second that my grandpa and my dad and my uncle didn't love each other, because they did. They had a relationship with each other. They had a bond.

I didn't have that bond with my dad, and it worried me. What if I couldn't live up to his impossibly high standards? What if I wasn't good enough? Then what would I do?

My dad has always been a hard man to say no to, but there was another reason that helped persuade me.

In early 2000, while I was out fishing, Lisa gave birth to our second daughter, Trinidy, whom we named after the Trinity Islands in Alaska. With a bigger family, I needed a bigger income.

"OK," I said, "let's do it."

"All right, let's find ourselves a boat."

So we started looking around.

Unlike me, my dad has always been careful with money. I knew he would do an exhaustive search and weigh the pros and cons of each boat with a cost/benefits analysis. He wouldn't sign his name to something without doing his homework.

I figured we'd be looking for a while, and I'd use that time to get some more experience as a captain. Experience, I would learn, that I desperately needed.

—◦—

It took us about a year to find the *Seabrooke.* Ironically she was one of the first boats we looked at.

The boat was pretty run down, but the hull was solid and the equipment was operational. She just needed some love and attention, but the owner, Stu Ferris, wanted more money than we were prepared to shell out.

So we passed on it.

We kept searching, though, and we spent about a year taking a look at a few different boats, but none of them had the kind of platform we were looking for.

Like everyone else in the industry, Stu was waiting for the fisheries to switch to a more regulated program. He was getting up there in years and couldn't afford to wait forever. He was having financial troubles, and the boat wasn't producing. He was barely hanging on. He was going to lose the boat if he didn't bring in some partners. He figured it was better to retain part of the boat than to lose it all.

Once he lowered his asking price to one million dollars, the *Seabrooke* became the most attractive option for the money that we had.

I was making about seventy-five grand a year running the *Vixen.* That was more money than I'd ever made in my life, but it was nothing compared to the million dollar debt we'd have to take on.

My dad's situation wasn't much better. He was making the same amount that I was, give or take, but he was putting at least half of that back into his farm. He was still trying to keep it going. He had the worst luck with that farm. It didn't matter how hard he worked, it was like Mother Nature held a grudge against him. She would split his crop, burn his fields, and freeze him out, and there wasn't a thing he could do about it. He had one of the nicest, cleanest orchards in the

Walla Walla Valley, but bad luck and worse weather would screw him over every year.

If it had been me, I would have gone into something else, but he had a passion for farming. He believed in it.

Things were very, very tight. We weren't sure if we would be able swing the deal. The only way we could make it work was to have Stu carry the note. We couldn't go to a bank. The banks wouldn't touch us because our seasons were short. From their perspective we were a bad investment, too risky. We made an arrangement with Stu so that each year we'd take on a bigger percentage of the boat until we owned it outright and Stu was free and clear.

The lawyers drew up a contract, and we signed our names on the dotted line. Just like that, I was a part owner of a million dollar boat.

I'd like to tell you I was overjoyed, but I wasn't. Far from it. My first thought was, *What the hell have I done?*

The *Seabrooke* wasn't a pretty boat, but my dad insisted that all she needed was a little TLC. Isn't that what they always say about junker cars and difficult women (or, as my wife tells me, difficult men)?

We officially became part owners of the boat in December 2000. Her keel was laid in 1978, which made her twenty-two years old, but she looked much older than that. She was 109 feet long and 30 feet across at the beam. That was unusually wide, but it gave the *Seabrooke* a stable platform. This was by design. Even though she was built in California, she was made to handle the rough waters of the Bering Sea.

The boat was a shambles when we came in. The owner hadn't been putting any money into it, but now it was ours. With opilio season coming in January, we needed to get her ready in a hurry.

My dad was still running the *Lady Alaska* and I was the skipper on the *Vixen* when we took over the *Seabrooke*. A guy named Tony Laura had been the captain of the *Seabrooke,* and we let him run the boat for one more season. We didn't want to kick the guy off the *Seabrooke*, because we knew he wouldn't be able to get another boat on such short notice. Besides, my dad and I had both made commitments to the boats we were running. So we decided to let Tony captain the boat for that one last ride.

We got the *Seabrooke* into reasonably good shape for opilio season, and then my dad and I went back to our boats. For that season, not only would we be competing with each other, we'd also be going up against our own boat.

Then disaster struck.

Shortly after the *Seabrooke* left Kodiak for Dutch Harbor, one of the main engines—a Caterpillar 675 HP—went down, and Tony couldn't bring it back online. He called us up to tell us that one of his mains was completely shot and that he would miss the season. This was a major catastrophe.

My dad and I were pretty much tapped out. We'd spent every last dime we had getting the boat ready to go catch opies. We couldn't afford to replace the main engine. It's not like the mechanics would replace it for free in the hopes that we'd make enough money to cover the bill, and it was a big bill—twenty-five grand. If we wanted to get the engine fixed, we had to pay upfront.

So we put it on a credit card—all twenty-five thousand.

Talk about an uneasy feeling. Here we were brand-new owners of a boat, and we were already adding to our mountain of debt. But there was nothing else we could do. Fishing vessels make their money by fishing. We couldn't afford to miss a season while we dicked around with the engine. We needed to go fishing.

Tony made it out for the season. I think he was a day late, but he had an okay catch. It wasn't the best season, but the boat made a little bit of money and we were able to pay off our twenty-five-thousand-dollar credit card bill. As soon as we got the boat going again, we turned around and went right back out at the start of cod fishing season. We worked that boat as much as we could just to pay off our bills and make ends meet.

—

When it was my turn to take the helm of the *Seabrooke*, I made a grim discovery: Being an owner didn't agree with me.

Don't get me wrong. I liked everything about being part owner of my own boat—bringing the boat back to respectability, being my own boss, calling my own shots. Unfortunately the boat didn't like *me*.

In 2001 I took the *Seabrooke* out cod fishing for the first time and ran into some bad weather. I started to feel a little ill as soon as we got out in open water.

At first I thought it must have been food poisoning, but no one else had gotten sick and we'd all eaten the same food, so we were able to rule that out pretty quickly. Was the *Seabrooke* making me seasick?

It certainly felt that way. I was nauseous. I was puking. I could barely sit in my chair. I was definitely seasick.

When I was a greenhorn, I got seasick a lot. But then I got over it and that was that. When it came back, I didn't put on a patch or take any medicine for it, I just rode it out. As soon as the weather cleared, I felt a lot better. To this day whenever the boat starts to buck into bad weather, I get sick. I don't throw up anymore, but I get nauseous and a nasty headache. It's the strangest thing. I can jump on any other boat in the fleet and fish through storms, and it doesn't phase me in the least. But for whatever reason the motion this one makes in rough weather doesn't sit well with me. Trough ride. Side to side. No problem. But straight into the weather gets me every single time.

It wasn't just the weather that gave me headaches. The boat was so run down it felt like it was falling apart. Usually when you buy a boat, you take it into the shipyard and see what you've got. Not that I'd ever done it before, but that was the protocol with boats that I'd worked on in the past when it changed owners. We'd break it down, clean it out, and replace all the fluids—at a minimum. If we had more time we'd replace all the parts that were corroded or breaking down so that everything was tip-top. If we had done that with the *Seabrooke,* we would have ended up with a brand-new boat. *Everything* needed to be replaced.

This wasn't a surprise. We knew what we were getting into—or at least we thought we did. After all, the main reason why the owner lowered his price was because he didn't have the money to maintain the boat. It wasn't that he was thrashing it, he just didn't have the money to put back into it. It takes money to make money. That's especially true in fishing; if you don't have the money to keep the boat ship-shape, it ends up costing you more in the long run, as you constantly have issues.

Every time we made a trip, something would break down. Whenever we came back into port to offload the crab, while the rest of the crew was sleeping, I'd be down in the engine room with the engineer trying to get the thing back together again just so we could pick up the gear we'd left soaking in the water. Then the process would start all over again.

We were constantly fixing something. I never got any sleep because the maintenance was so astronomically time-consuming. Worst of all, the majority of my repairs were short-term solutions. I was just trying to keep the boat bandaged together in order to keep fishing.

The boat's previous captains didn't have ownership in it, so they didn't take care of it. They didn't put in the extra effort to keep the boat running. Instead of trying to fix things themselves, they'd wait for the owner to order the parts they needed so that the mechanics in Dutch Harbor could fix it. There was no incentive for them. It wasn't their problem.

When my dad and I took over, we inherited every single one of those problems. It took everything we had just to keep the boat afloat.

For the first few years that we owned the *Seabrooke,* I worked the boat only during cod season in the spring and fall. When I went back to the *Vixen* for crab season, my dad took over on the *Seabrooke.*

The *Seabrooke* coming to the entrance channel of Dutch Harbor to unload crab, and leaving Dutch Harbor for the crab fishing grounds.

After a season on the *Vixen,* which was a brand-new, top-of-the-line boat, I hated coming back to the bucket of bolts that was the *Seabrooke.* It was like renting a Ferrari and coming back to a beat-up farm rig. In fact that's exactly what it felt like sometimes, like I was back on the farm doing double duty just to keep it going.

The boat had a lot of potential, but we were up to our ears in debt. All the repairs ate into our profits. However, it would be misleading to blame all of the repairs on the boat. There was some operator error involved as well.

During my first cod season, I had to fish on one engine, because one of the mains was down. There was no repairing it. It had to be replaced. Losing the engine cut my mobility in half, which was really frustrating. It was like driving down a seventy-mile-per-hour freeway and only being able to go thirty-five. It was slow going, but we somehow made it through the season.

Well, almost made it.

We'd finished up the cod season, and I was taking the boat to Kodiak to go tendering for the summer. I was able to deal with it pretty well—all things considered—until we ran into a storm. I just didn't have the maneuverability to fight through it. So I tried to take a short cut by taking the false pass off Unimak Island, the largest of all the Aleutian Islands.

It was called "false pass" because it was really wide on the surface—about three miles—but there was only about one hundred feet of water deep enough to drive through. And to make matters worse, it S-curved through the pass. It looked like a straight shot, but that wasn't the case at all. There were a lot of sandbars, which were always moving around.

It was risky, but I'd run other boats through it. Normally it saved a day's travel. This time the savings would be even greater because we were running at half-speed. So instead of going all the way around the island, we cut right through.

The *Seabrooke* loaded with cod pots in Dutch Harbor, Alaska, getting ready to start fishing cod.

Unfortunately I didn't take my compromised maneuverability into consideration, nor did I compensate for the *Seabrooke*'s extra depth. I was used to boats that had a shallow draft of six to eight feet. The *Seabrooke* was sixteen feet at the stern.

I went barreling through the pass, half-speed ahead. I was off track by a little bit but not too much—or so I thought. There was a big ground-swell of about four feet. The boat went up, and when it came down the keel hit bottom and made a sound you never, ever want to hear. We came up and went back down and hit again. We hit bottom three times before the swell bounced us into deeper water and we were through.

But at what price?

I was terrified I'd broken the rudder, but it seemed to be responding okay. It was hard to tell because we were running on one engine and our maneuverability was shot anyway. If all I'd damaged was the rudder shield, I'd be in pretty good shape. That's what its job was: to protect the rudder. But if the rudder itself had been damaged, I was hopelessly, royally screwed.

I chastised myself for being too cocky, too confident, too aggressive. Bottom line: I didn't know the boat as well as I should have. This was 100 percent on me.

Once we got to Bristol Bay, I sent a diver down to assess the damage. The diver came back with good news.

"Everything looks okay. She's a little bit bent, but it looks all right."

This was the best news I'd gotten all year. I ordered new bearings for the rudder and got them sent up to Dutch Harbor.

At the end of the summer, right before king crab season, my dad and I were getting the boat ready. As soon as the boat was prepared, I'd head over to the *Vixen* and my dad would take over the *Seabrooke*. When I put the boat in reverse, I heard a loud bang come from the stern. It sounded like we'd been rear-ended by a truck.

"What the hell was that?" I asked.

"I hope that wasn't the rudder," my dad said.

It turned out his intuition was right. That sound we'd heard was the rudder breaking. Now we were really in a bind. The season was about to start and we had a busted rudder. We called up the welders for an emergency job and got them in the water right away. They spent all night in the water working on the rudder.

As dawn broke over Dutch Harbor, we tested the rudder and a familiar sound erupted from back aft. This time we knew what it was. The weld hadn't held. The rudder had broken *again*. Now king crab season, our most profitable season, was getting underway without the *Seabrooke*.

We put the divers back in the water and had them jerk out the rudder and haul it up onto the dock where it could be repaired properly. After another day we were able to get it fixed and reinstalled. The total cost of the repair was ten grand.

This time the weld held and lasted throughout king crab season. But shortly after I took the boat out to go cod fishing, it busted again. It was too late to go back to Kodiak, and I wasn't going to shell out another ten grand for a half-ass repair job. So I took the rudder out and fished the rest of the season on one rudder. This made the *Seabrooke* almost impossible to turn. It could turn one way, but not the other. Imagine driving a car with a steering wheel that will turn left, but not right.

To say it was a challenging season would be an understatement—it was a huge pain in the ass. When we were hauling gear I had to be meticulous about how I positioned the boat as I approached the buoys, because I could only correct to starboard, never port. With just one rudder it's easy to get off track a little bit, and in bad weather, that can be dangerous.

To protect the crew working out on deck from big waves, I needed to be able to turn the boat quickly in order to stay out of the worst of the weather. With only one rudder, I couldn't do that. I couldn't protect the crew. So the crew ended up getting exposed, and we had a few wipeouts. But it's the waves you can't see that get you.

Waves come at you from all angles. They can hit you sideways, they can hit you straight on, or they hit you on the stern. The most dangerous waves are those that quarter you. (If the boat is a clock, your after starboard quarter is at 4:30 and your after port quarter is at 7:30.) A quartering wave simultaneously pushes the boat forward and sideways at an angle. That's called breeching the boat. When you breech, you're momentum accelerates faster than the boat can handle and you lose control of the vessel. There's literally nothing you can do.

I was running from a storm on one main. I didn't have any speed, so I was getting quartered left and right. I took a wave that breeched the boat when a second wave came over the top of us and totally submerged the deck. I couldn't see the sorting table. I couldn't see the rails. The deck was flooded with three and a half feet of solid water. We were underwater and squandering down. I could feel the boat fluttering, trying to come up.

She was doing everything she could, but we kept getting hit by wave after wave that pushed us farther and farther down.

This is it, I thought. This is how it ends.

Finally we caught a break between the waves, and the boat was able to rise up a little, get some of that water off her deck, and pop back up to the surface. It was the closest call I'd ever had. We were right there on the edge. If the *Seabrooke* had had a little less flotation and freeboard, we would have gone down. After that night I understood how a boat could sink in thirty seconds, and I never wanted to experience it again.

I'd never been scared of the sea—until that night. That scared the shit out of me.

When the season was over, we took the *Seabrooke* down to the shipyard and replaced the rudder with a brand-new one.

My misadventures with the rudder ended up costing us a grand total of one hundred thousand dollars. A one-hundred-thousand-dollar screw up.

I couldn't wait to get back to the *Vixen*, where I could count on the boat not breaking down on me every single trip, but my dad wanted me to come over to the *Seabrooke* full-time and work as a relief captain.

I was reluctant to go. In my mind I was well on my way to being an experienced captain. I didn't have the years under my belt, but I was getting there.

To my dad's way of thinking, I was still wet behind the ears. I wasn't sure if I could go back to being his apprentice. I wanted to prove to him that I was just as good a fisherman as he was. If he were looking over my shoulder on the *Seabrooke*, I wouldn't get that opportunity. The *Vixen* was the only boat I could captain to make a name for myself.

Finally, after we'd had the *Seabrooke* a couple of years, my dad gave me an ultimatum.

"If you don't come over now," he warned me, "I'm going to get somebody else."

The timing was less than ideal, because—as usual—I was up to my eyeballs in bills, but my dad didn't want to hear it.

"If you don't come, somebody else is going to be sitting in the captain's chair."

That was all the motivation I needed.

CHAPTER TEN

THE CODFATHERS

I HAVE A CONFESSION TO MAKE. THE SECRET TO THE *SEABROOKE*'S SUCCESS has nothing to do with my skill as a crab fisherman. I owe everything I have today to cod. If it weren't for cod fishing, we would have lost the *Seabrooke* a long time ago.

When my dad and I got the boat, we didn't have a lot of money. So we had to get creative and find ways to make some extra cash. We started to look around, and that's when we first began to consider the cod fishery as a serious opportunity for the *Seabrooke*.

My dad and I had always fished cod, but we were the outliers. Back in the day there were very few crab fishermen in the cod fishing community. Why? Because cod fishing required three times the effort for a fraction of the payday. Crabbers looked down at cod. Their attitude was, "Why fish cod for nineteen cents a pound when I get anywhere from two to five bucks a pound for crab? Why would I want to work my ass off for three or four months and make ten grand when I can make fifty grand in four to six weeks?"

They wouldn't do it. The crews wanted no part of it, and neither did the skippers.

Crab fishermen are proud people. They risk their lives and work their asses off every time they go out to sea. They didn't want to jeopardize their livelihood fishing for what they viewed as a substandard wage. They thought cod fishing was beneath them. Once that idea took hold, it was hard to get them to change their minds.

I understood where the crabbers were coming from. After I went cod fishing for the first time on the *Beauty Bay*, I swore I'd never do it again. It nearly killed me. But I got my first opportunity as a captain by fishing for cod on the *Amatuli*. Cod had been both good and bad to me.

Sometimes pride can get in the way of a golden opportunity. I wasn't too proud to work for nineteen cents a pound. Maybe it was because I was new to the fishing industry, but I wasn't about to overlook a fishery just because others guys weren't doing it. I've never been one to follow the herd. If other fishermen weren't making any money at it, that didn't mean I couldn't. Money was money, and I wasn't about to leave some on the table for someone else to get just because I didn't want to work hard. That's one thing about the Campbells. We are not afraid of hard work.

With cod it's all about volume. An average haul for a three-day trip was fifty thousand to seventy thousand pounds. There was no money in it. Boats were going backward.

My dad and I thought that if we could get that number up to one hundred thousand pounds per trip, we could make it a profitable fishery. We didn't care what others thought or had to say about it. We had to go out and try it for ourselves.

That said, we needed to make some changes. The *Seabrooke* was barely up to the demands of crab fishing, much less the extreme rigors of cod fishing. If we were going to make a go of it, my dad and I needed to fish smarter and stretch those dollars. We had to figure out a way to make money on every single trip.

We started with the boat. When we finally got the *Seabrooke* in the shipyard that summer, we decided to convert the boat to a cod boat. We set the deck up for cod fishing so that the crew could work more efficiently and haul more gear. Cod fishermen use the same pots that crabbers use, they're just rigged a little differently. It wasn't a complete changeover. We could still fish for crab, but since cod was the more difficult fishery, we wanted to design the deck for maximum efficiency during cod season.

But it was more than a matter of refurbishing the deck. We had to completely change our mind-set, too. We knew we couldn't go cod fishing with a crabber's mentality. Crabbers stack and move. A crab fisherman brings the pot aboard, gets the product out, stacks the gear, and moves on to the next location. Cod fishermen turn their gear over, meaning as soon as they get the product out, they put the pot right back in the water. It's a faster way to fish.

Most crabbers were hauling 100 to 150 pots per day. My dad always shot for 200 to 250. But cod fishermen haul 300 to 350 pots every single day. That's an incredible pace. It makes crab fishing seem slow by comparison.

We weren't a crab boat anymore. We were a cod boat that went crab fishing between cod seasons. Crabbing became our bonus trip, a way to make extra money. That's the approach we had to take if we wanted to be successful in the cod fishery: 100 percent commitment.

When we first got into it, cod fishing was something we could do six to eight months out of the year. There were two seasons—one in winter and the other in fall—and each one lasted from three to four months. The crab quotas were so low during those last days of the open-access era that most crabbers were only fishing a couple months out of the year and their boats were sitting idle the rest of the time. My dad and I didn't go a million dollars into debt so we could park the boat at the pier for most of the year.

We put the *Seabrooke* to work.

Cod fishing is a grind. Don't let anybody tell you different. The biggest difference between crab fishing and cod fishing is that with crab you've got live animals in your tank, but with cod you're dealing with a dead product.

When I went crab fishing, if I was having a miserable trip and wasn't catching any crab, I could take as long as I needed to fill up. If it took ten days, twenty days, so be it. Of course if I took that long I would get my ass handed to me while the other fishermen gobbled up the quota. The point is, I didn't have to worry about spoilage because the crab were alive and well in a temperature-regulated saltwater tank.

With cod it was a totally different story. The clock started ticking as soon as I caught my first fish. Once I caught it, killed it, and put it in the tank, I had seventy-two hours to deliver that fish to the processor. Even though the cod were kept chilled, they started to deteriorate shortly after they were killed. It didn't mater if I had ten thousand pounds or one hundred thousand pounds, if I didn't get that fish to the dock within seventy-two hours, I wouldn't be able to sell it.

Seventy-two hours. Three days. That was it. If I had a bad second day, I didn't get an extra twenty-four hours on the back end. When we went cod fishing, we had to make every hour count.

That was the main reason why so few crabbers got into cod fishing. Not only was it grueling as hell, it was easy to lose money. You couldn't afford to have a bad day when you went cod fishing. It could ruin your trip. String a few bad trips together and all of a sudden you were going backward and losing money and your season was shot.

The pressure was nonstop. I fished through storms. I fished through current systems. If a fifty-knot blow came up, I fished through that fifty-knot blow. The margin for error was so small that it made me a better crab fisherman.

"You think you're such a great fisherman?" I used to say to crabbers who liked to brag about how badass they were. "You ought to come cod fishing with me. Then we'll see how tough you are."

No one ever took me up on that offer. They viewed it as a high-risk/low-reward proposition. It wasn't that they were reluctant to put in the hard work or afraid that their boat might lose money. They were afraid of what it would do to their reputation. What if they couldn't hack it?

A tank full of king crab on the *Seabrooke.*

That's why so many crab fishermen stuck to crabbing. They were afraid to fail. They just didn't want any part of it.

But once the price of cod started to creep up, and my crew and I were making some serious money, you better believe they paid attention. Before long, cod fishing really started to take off, and the *Seabrooke* was a big reason why.

Cod fishing was, and remains to this day, an open-access fishery. That means we're not only racing against the clock, we're racing against the rest of the fleet to catch more product. It's a derby-style fishery.

In the beginning, cod fishing lasted anywhere from two to four months depending on the season, but as we became more efficient, the seasons got shorter and shorter.

For instance, last year I got five trips out of a twenty-day season. That's it. But I caught in three weeks what used to take me three months. That's almost half a million dollars worth of cod. When you're pulling in twenty-five thousand dollars a day, you can't afford to miss any days. Every day is crucial.

Now that people know there's real money to be made, more boats have gotten involved. It's become a more aggressive fishery. We're catching the quota more quickly every year, because every year we get a bit more efficient. It's the evolution of the industry, and as with every other business, you either evolve with the times or go extinct.

Crab fishing is a lot cleaner than cod fishing—that's another reason why a lot of captains and crews preferred crab to cod. There's a lot less mess with crab.

In cod fishing we use crab pots that have been modified for cod. We haul the pots, open the door, and the fish tumble down onto the sorting table that we use for crab, except we don't sort the cod—we kill them.

When the cod comes on board the vessel, the first thing we have to do is bleed the fish. A cod's heart is in its throat. We cut their throats so that they'll bleed out. It sounds brutal, but when you go to the supermarket to buy white fish, you expect white fish. If we didn't bleed the fish, those nice white fillets would have a red tint. Nobody wants that. So when the fish comes in to be processed, the quality has to be perfect.

We cut their heart so that it pumps the blood out of the fish. The guys on deck take a knife and poke them right in the jugular with a stabbing motion. They stab down into the sweet spot, and they move the fish into the tank. So instead of grabbing the fish with their hands, the guys use knives to poke, pin, and move the fish into the tank.

It's not pretty. It's messy. It's gory. The deck is awash with blood. The first time I saw it, I thought it looked like something out of a horror movie. As soon as we start bleeding the fish, every bird in a one-hundred-mile radius is flying around the boat, whipped into a frenzy. I don't know where they come from. We might see one or two seagulls while we're traveling, but as soon as we start fishing, the air is filled with thousands of birds. It's amazing.

It only takes a few minutes for the cod to bleed out. The water in the tank is constantly circulating with refrigerated water that pulls the temperature of the fish down and preserves the meat. That's when the clock starts ticking, and we have seventy-two hours to catch as much fish as we can, get back to the cannery, and offload the product. Then we go back out again. We do that over and over again until the quota is met.

The size and weight of the cod vary quite a bit. One time I caught a cod that weighed fifty pounds. That was a big, big fish. For the most part your average cod weighs eight pounds and is thirty inches long. When the fishing is good, we can fit about seventy-five fish in a pot. Average fishing you'll get about half that many. Anything less is pretty shitty.

Ultimately I don't care how many fish I catch. I care about how much they weigh. I shoot for three hundred pounds per pot. It doesn't matter if it's ten thirty-pounders or sixty five-pounders. I don't care, and neither do the processors.

Their one and only concern is the quality of the fish. As soon as we start offloading, a representative from the processor is there to evaluate the fish. White fish doesn't hold up very long. Once it starts to deteriorate, it turns to mush pretty quickly.

I've watched fisherman try to lie and cheat on their logs. "I was only out three days!"

But you can't cheat death.

If the quality is there, it's there. If the processor can't use it, they just grind it up. I've seen guys stand and watch as the processer grinds up one hundred thousand pounds of cod into meal.

The seventy-two-hour window isn't the only constraint we have to deal with. Once we start to fill up, we have to call the processor and schedule a delivery date. I can't just pull into Dutch Harbor anytime I want and expect a crew from the processor to be there to offload. It all has to be planned in advance. During the season, when there are hundreds of boats fishing for cod, if you miss your delivery window, it can be days before you get another one, and that can mean disaster.

In all my years of cod fishing, I've never missed a delivery date. Not once in my career. I've always made it. There were times when I cut it close to within an hour or two of my delivery date, but I always made it.

Missing a delivery date can have a deadly impact for crab fishermen, too. The longer you sit, the more crab you're going lose. That's called dead/loss: the percentage of crab that doesn't make it out of the tank alive. The last thing you want is crab dying in your tank, because when one goes, he takes his friends with him.

Crabs self-poison, meaning that when a crab dies, it releases a toxin that weakens the crab around him. One crab will poison four, and they'll poison twelve more, and so on. The cycle continues until you either get the live crab out of the tank or you have a boatload of dead crab that you can't sell and you can't eat. Crabs are voracious eaters who will eat anything dead or alive—except other crab. They won't turn on themselves because of the toxin. They can't eat their own. It's Mother Nature's way of stabilizing their species.

Cod, however, don't seem to mind. They eat their own all the time. They eat little cod. They eat pollock. They eat crab. During seasons when you have a lot more cod than crab, the cod will eat up all the baby crab. Cod are very aggressive fish—more aggressive than crab. Cod are intense. I would say cod are the most aggressive fish in the Bering Sea.

All cod do is eat and eat and eat. It's their purpose in life. One fifty-pound cod can munch through thousands of crab in a year. Luckily each female crab produces a million eggs.

Once we got the boat ready for cod, we had to find a crew that would come fishing with us. Not just once or twice but every season. In the beginning I had two crews: a crab crew and cod crew.

It took me a long time to find a crew who would do both. I had to convince guys that the cod fishery had potential. It wasn't a matter of skill, but faith. They needed to believe there was money to be made and that I had what it took to make it. Cod fishing was so demanding that the guys on deck had to know there was a reward for all the punishment they were enduring.

It was tough. I went through a lot of crew. There were times I'd lose a crewmember after every cod season. They'd work a cod season and a crab season back-to-back, and when they realized that one was so much harder and less lucrative than the other, they'd quit. There were seasons I'd have guys quit every trip, and I'd have to scramble to find new guys hanging around the bars in Dutch Harbor.

Not all of these guys were quality fisherman. Some of them were just guys who'd come up to Alaska looking for an adventure and some quick cash and didn't count on having to work hard. Alaska attracts all kinds of oddballs, and not all of them are cut out to be fishermen.

Actually, some of them weren't cut out for anything. I had guys who were slobs and would go ten days without showering. I had guys who were compulsive liars. I had guys who were thieves.

There's nothing worse than a thief on a boat. Every now and again I'd get a guy who thought he could get away with taking something that didn't belong to him, which I never really understood. You're out in the middle of the Bering Sea, there are six or seven guys on the boat, and something comes up missing. Obviously *somebody* took it, and ninety-nine times out of a hundred it's the new guy. Who do you think is going to catch the blame? It's not rocket science. Then again, thieves usually aren't the sharpest knives in the drawer.

One time I needed a guy so we could finish out the season. I was desperate, so I hired a cannery processor. After a couple of days, a laptop disappeared. Then an iPod went missing. I never confront a guy when we're at sea, because there's no telling what he'll do.

I've never had a member of my crew lose it, but there have been times when a guy will get some bad news when we come in to deliver. He'll find out that his girlfriend is sleeping around or his old lady wants to break up, and instead of talking about it and getting it off his chest, he'll brood over it and let it fester in his mind. Then, when we're back out at the grounds, something will set him off and he'll lose his temper over something that normally wouldn't bother him.

I've heard stories about guys having to be quarantined to their staterooms after flipping out on the boat. They demand to be taken off the boat and then threaten bodily harm to the skipper if he doesn't turn the boat around.

I've never personally experienced that drama, but I've had guys quit on me, and it's always uncomfortable. When they get done ranting and raving about whatever it is that got them upset, I calmly explain to them that we're not stopping the trip just because they want to go home.

"If you had business that needed to be taken care of," I'd say, "you should have stayed in town."

Usually they calm down, but sometimes they don't, and you have the thought in the back of your mind that they might try something. Is this guy going to try to kill me in my sleep?

That's why I always wait until we're on land to confront a thief. On this particular occasion, with the guy who stole a laptop and an iPod, I waited until we were at the airport getting ready to fly out.

He probably thought he was going to get away with it—until I confronted him.

"Here's the deal," I said. "You took our stuff. Give it back, and I won't press charges."

The guy got upset. "I didn't take nothing!"

My guys were pissed. There were four of us and only one of him. They grabbed his backpack and opened it up. Sure enough, there was the laptop, there was the iPod.

I gave the guy two options.

"Stay in Dutch Harbor and get arrested or leave town and never come back."

I felt like a deputy in a Wild West movie. Needless to say, he took the second option.

Guys come to Dutch Harbor from all over the world. A lot of them have checkered pasts. All of them are looking for a second chance. People come to Alaska to get away from the mess they've made down below. It's always been that way. Some of them can change their ways and start over and some can't. We're willing to give just about anybody who's willing to work hard a chance, but don't steal from us.

We're a family out here. A fishing family. This is our home. We don't lock up our stuff. We leave our wallets and phones and computers out just like we do at home. If you need a hundred bucks, ask for a hundred bucks. If you need a computer, ask for a computer. We trust each other with our lives every single day. Without that trust, we're not a crew. We're just a bunch of guys in a boat.

One of my best guys was a fisherman named Moose, but he didn't start out being one of my best. Prior to cod fishing, Moose was just an average deckhand. In appearance, however, he was anything but average. A six-foot, six-inch 280-pound black man is going to stand out in Dutch Harbor. He was your typical gentle giant. Everybody loved Moose. We'd come into town and everywhere he went people would stop and say hello. He'd go around to all the bars, and shoot the shit with all the fishermen. People really enjoyed his company. He was an all-around good guy.

But on deck he could be frustrating. He was a big, powerful man with incredible size and strength, but he didn't know how to use it. He was such a big man it seemed as though he was embarrassed by it. I felt like he just wanted to fit in and be a normal guy.

One night we were fishing through a nasty storm. The rest of the fleet had already called it quits and gone back to Dutch Harbor. We were behind schedule, and I was trying to make up for lost time. We were under the gun, and I was feeling the pressure.

Moose was down on the deck throwing the hook. The thrower's job is to take a five-pound grappling hook attached to a hundred feet of line and throw it as far as he can so he can snag the buoy, get the pot line in the block, and haul the gear. It's a hard job, and Moose was half-assing it, just going through the motions with zero sense of urgency. We couldn't afford to miss a pot. Every minute mattered. I got on the loud hailer and chewed him out.

"Dude, you're the strongest guy on the boat, but you throw the hook like a pussy!"

That made Moose mad. It really pissed him off, and he took it out on the hook. He started tightlining that hook with every throw, meaning he was throwing the hook one hundred feet, which was just about the length of the boat. Not many people can do that. Moose could, but he didn't know his own strength. Here he was making these phenomenal shots, pulling the line as hard as he could and getting it in the block in no time at all. In other words, he was doing what he should have been doing all along.

At the end of the string, he came up to me. It would have been way out of character for him to make trouble, but I'd never seen him angry before either.

"Thank you," he said.

"For what?" I asked, somewhat relieved.

"No one else has ever pushed me like that."

I didn't need to say anything else. The light had clicked on for him. He got it. After that night he was a great deckhand and the best hook thrower I had on the boat. I haven't had a better one since.

Of course not everyone who comes up to Dutch Harbor can make that kind of transformation. I had one guy who was a good worker, but every time we went out he got hurt. Every single time. He was another guy with a shady past, but he'd cleaned himself up and was a great deckhand. He just couldn't stay out of harm's way.

I tried to protect him as best I could. He'd been to culinary school, so I made him our cook. He had a pet peeve about keeping the counters clean, which I guess is something they teach you in culinary school—always keep your work area clean. Good luck with that on a fishing boat.

One time he was making dinner and bitching to me about how the crew always made a mess of his counters, and the whole time he had his hand down his pants.

"Are you talking to me about sanitation?"

"What do you mean?"

"You just scratched your balls!"

"I didn't do that!"

"Dude, I just watched you!"

He denied it all through dinner, and I made myself a peanut butter sandwich that night.

              —◆—

We had some rough seasons. Bad weather. Horrible ice. Low quotas. All the things that can break a person's will, but I never stopped believing that we could make a profit in the cod fishery.

Even when we started making decent money, some guys just didn't want to put in the work. It took me several years to get a crew of true believers together, but once I got the right crew, everything fell into place. We started making good money. Not just once or twice, but every single season.

When the price of cod went up to thirty, forty, and fifty cents a pound, my guys were making big bucks. They were getting huge checks at a time when the crab industry was going through some hard times. It was the twilight of the open-access era, and times were tough—except for us. We were still making money during crab season, but cod was catching up. More and more every season. When it finally happened and my crew made more money cod fishing than they did crab fishing, they were completely hooked.

Then everybody wanted to come cod fishing with us. Deckhands were begging for a spot on the *Seabrooke*.

I wasn't the only fisherman making a killing in the cod fishery. There were a handful of other boats that had figured out how to turn a profit with cod. We were known as the Codfathers. We were the founding fathers of the cod fishery, so to speak. That's when I started to slide out from under my father's shadow and stand out among fishermen. My dad didn't go cod fishing. That was all me. I was finally making a name for myself as a fisherman.

I had gone into a fishery that no one else wanted to fish and figured it out on my own. I refined my methods and made it efficient so that I could get those numbers every time. Every fisherman had fluke seasons—good and bad. Consistency was everything, but that's not what I was striving for. I wanted to be the best.

My goal every time we went out was to be the number one boat. *Almost* didn't cut it. Nor was I satisfied with the top slot. I wanted the *Seabrooke* to be so far ahead of everyone else that they had no chance of catching up, much less beating me.

This was nearly impossible, of course. There were times we did it and there were times when we weren't even close. We had our share of bad seasons just like everyone else. But during cod season, no one could touch us. How could they? We were the Codfathers! That was our attitude. We were the best, and we were unbeatable. Even when I came in second, third, or further down the line, I was relentless in my desire to succeed.

I set the bar extremely high. I was bringing in three to four times more than what the rest of the fleet was catching. As the boat got better at cod fishing, our profits increased. Word started to get out about cod. Captains and crewmembers started taking an interest in the fishery. Pretty soon we had thirty-five to forty boats that were out cod fishing. When that happened, we started to catch the quota a lot faster. Every season we'd see more boats out there. Cod season went from four months to three months to two. We kept getting more and more efficient at it. Now a season only lasts a few weeks.

For many years I couldn't get past the 300,000-pound mark for a single trip. That was the unbreakable number. I came close. I had 296 one time, 285 another. It seemed impossible, but once one fisherman did it, others followed suit. The highliners started catching 330,000, 345,000, and 360,000 pounds. In 2013 we broke 300,000 pounds four out of our five trips, and the first trip we had 270,000.

At the time of this writing, I can fill up my tanks in three days with 335,000 pounds. It's not easy, and I don't do it every time. I might do it once, maybe twice a season, but it can be done. Fifteen years ago that would have been a record breaker. That's how quickly things have changed.

Even after we mastered the fishery, the biggest challenge was time. It became a chess match. For instance, some guys would try to fish around storms. They'd go out to the grounds and wait for the weather to break before they'd put their first fish aboard, starting that seventy-two-hour clock.

I didn't care. I fished through storms. I fished through mechanical failures. I'd fish through anything if I was on the clock. Otherwise I was just wasting time, wasting fuel, and wasting bait. All for nothing.

To do that I didn't just push the crew, I destroyed them. I pushed and pushed and pushed. There were times when I moved all the clocks on the boat ahead to cheat the crew out of an hour of sleep so I could get them back on deck faster.

I never asked them to do anything I wouldn't do myself. I pushed myself beyond my limits. I was staying up two and three days at a time. You can't fish when you're asleep. If I was sleeping, someone else was out there catching. I slept on the way out and the way back.

When we were fishing the grounds farther from town, we'd have a six- to eight-hour drive back to Dutch Harbor. Usually I had one of the guys drive the boat while I rested and caught up on my sleep. Eight hours of sleep on a fishing boat is practically unheard of, but sometimes it was the only sleep I'd get the whole trip. When you go that long on so little sleep, weird things can happen.

Every fisherman I've ever known has had the boat dream. Captain, crewmember, it doesn't matter. We've all had it. The dream comes when I'm so exhausted from working around the clock that I don't know if I'm coming or going. In the dream, I'm cruising along when all of a sudden I see the beach approaching fast and I can't get the boat stopped in time. No matter what I do, I can't avoid it. Then all of a sudden, the boat kind of judders up onto the beach and I'm driving down the road like it's a car. The boat feels like it's on skates. I have control, but barely. It's loose, and I'm trying to dodge cars and make my way through intersections. Then along comes a building, and I'm about to hit it and—BOOM—I wake up.

In my head, I know it's not right. I know I shouldn't be able to drive the boat on the street, and yet in the dream it makes perfect sense. It's the boat dream.

You can ask any fisherman who's spent time at sea if he's ever had the boat dream, and he'll say, "Driving down the street with the boat? Oh yeah."

Every fisherman has had it.

Well, one time I was so exhausted that I almost made that dream come true.

We were finishing up a trip and steaming back to Dutch Harbor. We were about an hour outside of town, and I just couldn't keep my eyes open any more. I was rotating my guys so they were on for eight hours and off for four, but I didn't have a relief driver, so I had to stay up the whole time. I'd been awake for three days straight. Even though we were due at the dock in less than an hour, I decided to take a quick nap. I called one of my crewmembers up to the wheelhouse and had him drive the boat while I crashed.

"No matter what," I said, "wake me up in forty-five minutes."

There was nothing out of the ordinary about this. I often caught up on sleep during the drive back. I'd tell the guys how long I wanted to sleep, and they'd have a cup of coffee ready and waiting for me when they woke me up. Then they'd go down and get the lines ready to tie up, and I'd bring the boat in.

I lay down on the floor and went out like a light. An instant later my driver was hollering at me to get up again.

"We're almost there!"

I climbed back into the captain's chair and pointed the boat up the channel and sullenly drank my coffee. To get back to the cannery, we had to go around a big rock pile. There was a false lead where there was a dip in the bay before the rock pile. I had my computer screen with the navigation charts shrunk down and was going on memory. I knew when to make the turn. I'd made it a thousand times before.

Even though I was awake, I wasn't mentally alert. I was still out of it. After three days without sleep, I was a zombie. So even though I looked like I was awake, I was kind of slipping away with my eyes wide open. You know the expression, "The lights are on, but nobody's home"? That was me.

I made the turn, but instead of going around the rock pile I took it early and went up the false lead. I looked ahead and there was a beach in front of me. A small stretch of sand. I'd never seen this beach before, because it wasn't supposed to be there. *Huh.* And that's when it dawned on me that I was about to make a very big mistake.

I slammed the boat in reverse. Those V12 engines roared to life and shook the whole vessel. It damn near threw the guys on the bow off the boat. Thankfully the *Seabrooke* has a lot of horsepower and the boat slowly started to back up.

I didn't stop shaking until we were tied up at the dock.

The guys up on the bow told me they could see the bottom when they looked down. One more foot and I would have beached the boat and left a monument in front of Dutch Harbor for everybody to see. That's how close we came to being a statistic.

It happens every year. Fishermen run aground or beach their boats after going too long without sleep. They get so sleep deprived and fatigued that their bodies shut down on them. A five-minute nap turns into a half-hour snooze, and the next thing they know they're up on the beach. Only this time it isn't a dream—it's for real.

After I nearly crashed the *Seabrooke*, I instituted a new plan: When we're coming into port, a second crewmember has to be up in the wheelhouse with me to make sure I don't drift off into dreamland.

CHAPTER ELEVEN

A NEW ERA

In 2005 the Alaska Department of Fish and Game finally instituted a new program to regulate crab fishing in the state. The industry had been bracing for it for years, but it was still a shock when it came.

The open-access era was over.

The derby days were done.

A new era of crab fishing had begun.

The program was called the Individual Fishing Quota (IFQ). Every boat was issued a quota based on a number of factors, but the two main ones were the fishery's Total Allowable Catch (TAC) for the season and the boat's history in the fishery. The better your history, the bigger the quota.

Unfortunately this system put a serious hurt on a lot of the smaller boats. Their quota was so low that many owners weren't able to make enough to cover their expenses, much less turn a profit. About two thirds of the fleet went out of business.

Permanently.

The writing had been on the wall for a long time, but it had a devastating effect on Dutch Harbor. A lot of guys had to sell their boats, deckhands weren't able to find work, men and women who'd been in the community their whole lives had to pack up and go.

Crab fishing was always big business in Dutch Harbor. Hell, it was *the* business. But it was a business driven by a maverick spirit and individuals willing to take risks.

Not anymore.

When rationalization came into play, the industry lost a lot of its charm. The crab fleet went from *Cannonball Run* to NASCAR overnight.

That's not to say the new system was all bad. The fishermen were no longer competing with one another for every pound of crab. The IFQ determined how much we could catch, and we could go after it at our own pace. We still tried to outdo each other—we were fishermen after all—but we didn't have to push ourselves quite as hard, and that made for a safer, saner fishery. If one less fisherman ended up at the bottom of the sea, it was worth it. I miss the wild nights and winner-take-all spirit of those days, but there are some things you can't put a price tag on.

The *Seabrooke* came through all right. Would we have liked our numbers to have been higher? Sure. But mostly we were happy to have the opportunity to keep fishing. We didn't get into fishing because it was easy. We were grateful to be in the position we were in.

The new program had a lot of features that my dad and I liked. For instance, our IFQ was divided into shares, which could be bought, sold, or traded. Let's say my buddy's boat breaks down halfway through the season. Instead of losing his ass and going in the red for the season, he can call me up to see if I'm interested in buying the shares he can't use. He recoups some of his losses, and I get to add to my total quota—a win-win for both boats.

The new regulations had a calming effect on the entire industry. Prior to rationalization, the seasons were long and unpredictable. In January we fished opilio first and then, depending on whether it was a two-week season or a two-month season, we started cod fishing. Cod season would start anywhere from February to March and end a few months later. Again, the timing was always up in the air.

After that we'd take the boat back to Kodiak to get it ready for tendering. This usually involved a bit of light maintenance and quick repairs. Salmon tendering lasted from May until the end of August. As soon as we finished tendering, it was back up to the Bering Sea for cod fishing in the fall. That lasted until it was time to go king crab fishing, which was usually around the middle of October. But again, we never knew exactly how long we'd be out there. Then, as soon we were done crab fishing, we'd go right back into cod fishing.

The only time the boat was ever tied up was in December. We shut her down for a month so that everyone could go home and spend the holidays with their families.

One of the biggest changes from the old days is now fishermen can plan ahead a little better. In terms of logistics, we're much more efficient than we were before. The best part is now we have a better idea as to how long we'll be gone and when we'll be back. This eliminates a lot of the unknown, which can be really tough on families.

Lisa could testify to that. She was always on me to come home more often and stick around longer. The problem with that was, as a boat owner, I now had more responsibilities to take care of. If the *Seabrooke* needed repairs at the end of the season, I couldn't just jump on the plane, I had to make sure she was ready to go for her next trip. Lisa didn't see it that way. She wanted to know exactly when I was going to be home and for how long.

Fishing is never going to be an exact science. We can't predict the weather. We don't know for certain where the crab and cod will be and how much we'll be able to catch. We can't look into the future and see when parts will wear out and when the boat will break down. But rather than pull dates out of our asses or make them up on the spot, we can give our families a reliable estimate of when we'll be home. That makes a huge difference.

We still go home for Christmas, but everything else has changed. The seasons are a lot shorter, so the fisheries don't overlap. Now it's opilio in January, cod in March, salmon tendering in summer, cod in fall, and King crab in the winter. Every once in a while, they'll open a new fishery, like bairdi crab, and we'll go fish that, too.

In 2006 we got another crack at them. Bairdi are bigger than opies, but very hard to catch. It takes a lot of patience and it's very easy to run the boat backward. I'd fished it once before as a deckhand, but never as a skipper. It was a whole new ball game for me.

Bairdi are different than other kinds of crab in that they're very particular about their habitat: They love the mud. Normally we'd set our gear in long strings, but with bairdi we might drop only five or ten pots at a time. We had to find the right spot. If we dropped the pot in sand, we came up empty, but every time we hit pay dirt, so to speak, we'd haul in a full pot.

It took a lot of time. First I had to find the mud hole. Then I had to figure out how big the hole was. Once I had them zeroed in, I'd go to work. Even though they loved mud, they moved around quite a bit. There were times I'd catch some and think I'd found a mud hole, only to come up empty. It's a really frustrating fishery.

It was slow going at first. It took us a while to figure it out. It was an unconventional way to fish. Kind of like pothole fishing. The areas where we'd find them were tiny. The key was to jam as many pots in the mud hole as we could. Once I found a mud hole, I could stay on top of it until I caught them all.

Eventually I got the hang of it and we made a little bit of money. Not a lot, but a little. By the end of the season, I had my strategy down.

That's the way it is in the fishing industry. You have to pay your dues by trial and error, and it might take years to get another opportunity to use what you've learned.

There were plenty of trips where I had a hard time finding what I'd set out to catch. Whenever a fisherman says he came up empty, he doesn't mean the pot was empty. He means he didn't find what he was looking for, because the thing about pot fishing is you almost always catch *something*.

One time we caught a shark. A big, old ten-foot mud shark got stuck in the entrance to the pot. He thought he was going to snack on some cod

On the deck of the *Seabrooke* holding a bairdi crab while fishing off the Slime Banks on the Alaska Peninsula.

and got wedged in there. He was still alive when we brought him aboard, and he was thrashing around like crazy.

I got on the loud hailer, "Get that shark out of there!"

Nobody wanted to touch it. No one said anything, but I could see their faces. My crew was absolutely terrified. The shark was wedged in there pretty good, but there was no way they were getting near that thing.

The shark wiggled its way out of the pot, which made things even worse. Now we had a ten-foot shark flopping around the deck, and my crew was running around like a bunch of six-year-olds. Meanwhile I was laughing my ass off in the wheelhouse.

Finally someone was able to get a line around the tail fin and hoist that sucker off the deck and over the side. We didn't see too many sharks in the Bering Sea, but that was a mean one.

One of the more dangerous animals we encounter is the wolf eel. They chase after the cod and end up in the pot. They're about five feet of pure muscle with a twenty-four-inch head and some gnarly looking teeth. I've seen a wolf eel snap a wooden dowel in half with its bite.

They're sneaky little devils. If you get one in your cod pot, it'll blend right in. You'll be digging through the pots or the sorting table, and all of a sudden a wolf eel will jump out of a big pile of cod and snap at you. Those things will break your hand if they get a hold of it. They're pretty badass. You have to pay attention. You never know what will come tumbling out of those pots.

We catch a lot of octopus, too. It's not at all unusual to find a forty-, fifty-, or sixty-pound octopus in the pots. They can be tough to deal with. Those sixty-pounders can get up to eight feet long. They're not very aggressive, but they're stubborn. If you try to throw them over the side, they'll latch on to you and hold on for dear life. Or they'll grab onto the table with their tentacles, and it's next to impossible to get those damn things off. You have to wait until they start moving around. That's when you grab them and wrestle them off the boat, but it's not easy. They're fast and smart. They'll stick to your rain gear. Now all of a sudden you've got a forty- or fifty-pound octopus hanging off you, and it's easy to lose your balance. If you're not careful, they'll take you right over the side. Death by octopus is not something any fisherman wants written on his tombstone.

One time we had a greenhorn who was shit-your-pants scared of octopuses. We'd caught one and couldn't get it off the sorting table. It was a real big sucker, at least sixty pounds. Maybe bigger. Each of its tentacles looked like it weighed ten pounds.

As a prank I went down and cut off one of those tentacles. When you sever an octopus's tentacles, the limb will continue to move around for a bit. So I took the tentacle and slapped it over the greenhorn's shoulder. One part clamped down on his rain gear, and the other part wrapped around his neck.

The guy got so scared he started screaming like a two-year-old girl. He ripped his rain gear off and ran inside. The poor kid had pissed himself and had to go in to change his clothes.

Everyone laughed so hard and for so long, the whole production ground to a halt. All the deckhands were rolling around on the deck and couldn't get up. We felt bad, but only for a minute, because when he came back out on deck we all started up again.

The moral of the story: When you're a greenhorn, don't ever tell anyone you're afraid of something.

Sharks and eels and octopuses are all pretty commonplace. The strangest thing I ever caught was a canoe.

I don't know where it came from or how it got to the bottom of the Bering Sea, but somehow a canoe ended up in our pot. We were hauling gear when all of a sudden I saw this big, silver flashy thing in the pot. I thought it was some kind of monster fish, but once I got a closer look at it I could see that it was half of a canoe. The metal was all jagged and torn from where it had been ripped in half.

It was the weirdest thing. We were fishing way up to the north, where there was zero chance of a canoe ever being used out on the open water. I have no idea how it got there. The underwater current systems must have pushed it into the pot somehow. Every now and then we'd find a boot or a soda can, but a canoe?

That was interesting.

—◦—

When people think of storms at sea, they think of giant waves, but the wind can do just as much damage.

A few years ago I decided to install a wind meter on the *Seabrooke*. Incredibly, the boat had never had one. My crew was pretty good about estimating how hard the wind was blowing, but it caused a lot of arguments, mostly friendly. I finally broke down and bought a meter so we'd know once and for all.

Once we put it in, we took readings all the time just for the novelty of it. But we didn't have to wait long for it to come in handy.

We were trying to set some gear before a storm came in, but we didn't make it. I wasn't trying to fish through what was shaping up to be a big storm. I just wanted to get some of the pots off the boat to increase our stability.

The wind was blowing hard, and the waves were shipping tons of water onto the deck. I had one guy up on the stacks and another guy on the bow. We were taking so much water, I couldn't even see the guy up forward.

I checked my brand-new wind meter. It said eighty-five miles per hour with gusts up to eighty-seven. I nearly had a heart attack.

"Get off the bow! Get off the stack! Clear the deck!"

I had them come in. The guy on the bow came up to the wheelhouse.

"Jesus, how hard was that wind blowing?"

I told him it was eighty-seven, and he laughed.

"That's what it takes to call it on the *Seabrooke?* Eighty-seven-mile-per-hour winds?"

Since then it's been kind of a joke on the boat: Anything less than eighty-seven miles per hour and we're going fishing.

That storm was no joke. The wind blew so hard it picked up the water and lifted it onto the deck. It wasn't spray. A solid sheet of water slammed into the boat. I couldn't see anything through the wheelhouse windows. It was like looking through water, which was exactly what we were doing. And it was easy to imagine that was what it would look like if the boat went down. It gave me an uneasy feeling.

They say if you stay fishing in Alaska long enough, you'll see everything. Even a hurricane.

I was out fishing when the weather service sent out a warning for a storm with hurricane-force winds. The trouble with the weather service was that they were always calling for severe storms, and they were always downgraded.

Well, almost always.

All the boats went in, but I kept fishing. I didn't want to go back. I was having one of my best cod trips ever. Once in a lifetime kind of fishing. So I kept at it.

Sure enough the front side of the storm rolled in with sixty-mile-per-hour winds. It was a pretty good blow, but it sure as hell wasn't a hurricane. After eighteen hours the winds died down and it was as flat and calm as I'd ever seen the Bering Sea.

Those suckers, I thought. *They shouldn't have gone back to town.*

I continued to fish and had the grounds all to myself when my uncle called me.

"You better get headed back to town."

"What are you talking about?" I asked. "It's flat calm out here. The storm's over."

"We've got hundred-mile-an-hour winds here in town, and they're headed your way."

"Holy shit."

I left my gear in the water and took off right away. I was about four hours away from Dutch Harbor when I hit the storm. The wind came at me sideways and was blowing one hundred miles per hour. The wind was so strong we had a twelve-degree list.

I came up through the east channel to tie up at UniSea on wind power. I had the boat out of gear, and with the wind at my stern I was going seven miles an hour with no power. I was able to come around by putting the bow straight into the wind. I put the rudder hard over and walked the boat sideways into my uncle's boat, where we tied up.

While we were lashing on the lines, the wind ripped the roof right off the processing plant. Pallets were flying through the air. Workers were being teakettled down the dock.

We were the last boat to come in. If I had been on a boat that wasn't as stable as the *Seabrooke,* I probably wouldn't have made it. I was extremely lucky.

When it comes to bad weather, I've always been the one to push the envelope. I don't have a fear of it. It's exhilarating. I'm an adrenaline junkie. I'm not gonna lie. I always have been. Everyone always says to me, "If you weren't a fisherman, what would you be?" I probably would have been a stunt man or a racecar driver.

When I was young, I prayed for weather. The shittier the better. Nowadays, I do my best to avoid it. I have a greater respect for it. I have a better sense of when to call it a day. That's just maturity. The daredevil being forced out of me.

Being out in a storm is dangerous. There's a much greater likelihood of having something bad happen to the boat. And it's my boat.

The hurricane was hell on Dutch Harbor. It was the dead of winter, but those southerly winds warmed everything up. We had terrible mudslides. The mud came down, washed out the roads, and wiped out buildings. Entire warehouses were destroyed from mountainsides caving in.

That was an intense, crazy storm.

When it was over, I went back out to my gear. I had one hundred thousand pounds of cod in my pots. Every fish was dead. Every single one. The hurricane blew so hard and the seas were so big, that it churned up the bottom of the ocean and sandblasted the fish. They had no chance. They couldn't get away and were suffocated in the pots.

It was the most disheartening thing to haul our pots and dump out all that dead fish.

After the storm there were no more fish. They all left. Everything that survived went to deep water, and they didn't come back for the rest of the season.

Even the fish knew when to get the hell out of Dodge. They had more sense than I did.

In the Bering Sea if the wind doesn't get you, the below-freezing temperature will. Usually it's both.

Out on the Bering it's not just how hard the wind's blowing but where it's blowing from that matters to a fisherman. Let's say we had a relatively nice thirty degree Fahrenheit day. Common sense tells you that when the wind's blowing it will feel a bit colder.

If the wind is blowing out of the south, the temperature will drop only ten degrees or so, but if you get a northerly wind, with the wind chill it could be negative fifteen degrees. The north wind is ferociously cold, and when it starts blowing hard, it brings the temperature down in a hurry.

Not too long ago I was fishing way up north near the ice pack, and it was seven degrees. The wind was blowing out of the north at forty-five miles per hour, and the wind chill was negative fifty-two degrees.

You read that right. *Negative* fifty-two degrees.

We were hauling gear and dumped a load of crab on the sorting table. Within two minutes all of the crab dropped their legs and froze to death. It was like going into a blast freezer, and the crab didn't like it.

Neither did the boat.

All the water in the boat froze up, so we had no fresh water. All the lines were frozen, so we couldn't keep the boat warm. And the ice built up so quickly that getting it off was a full-time job. That was the coldest I'd ever seen it.

The *Seabrooke* is at is best when the weather is at its worst. It's not the biggest boat, the most efficient boat, or even the most reliable boat. There are plenty of boats that can haul more gear and carry more weight, but the *Seabrooke* is one of a kind.

For an average-size vessel, it's extra wide. I've seen boats that are twenty to sixty feet longer but not as wide. If you have a boat that's one hundred feet long and twenty-six feet wide, it's going to be very pitchy. There are limits to the weather that it can fish. Hundred-footers are typically not thirty feet wide, they're twenty-six to twenty-eight feet. That makes the *Seabrooke* a remarkably stable boat. You wouldn't think four feet would make that much of a difference, but it's crazy what this boat can do, especially in weather.

Fishing opilio in 2008 in the Bering Sea ice floes.

The *Seabrooke's* stability in big weather is a huge asset. The only trade-off is fuel efficiency. The boat was built in the 1970s when diesel fuel was fifteen cents a gallon. Nobody gave a shit about fuel economy back then. Most of the time the canneries would fuel up the boats for free. Back in 1978 the *Seabrooke* was the ultimate Alaska crab boat.

It's a little boat with the heart of a much bigger boat. There's nothing we can't do.

We can fish bigger weather, and we can take on more ice. Most boats get really heavy and lethargic when they start to ice up. That's when you know it's time to send the guys out to break up the ice.

Not the *Seabrooke*. This boat can pack a lot of ice. It just keeps taking it and taking it, which isn't necessarily a good thing. Neither my dad nor I know where the breaking point is on this boat. I have no idea how far I can push it. If we ever took the *Seabrooke* to its limit, I don't think we'd ever feel it. I think it would just get to the point where it can't take any more weight and go down. It's that skookum.

Shortly before rationalization, when my dad was running the *Seabrooke* for crab and I was running it for cod, if another boat became available for crab, I'd jump on it so that we could make some extra money, which we needed.

For one short king crab season, I was the skipper on the *Cape Caution*. It was exceptionally cold that year, and the ice pack came down to the fishing grounds. Dad was having a transmission problem and got a late start. By the time he got the boat fixed and headed back out, we had a big storm come up, a northwest storm, and the *Seabrooke* bore the brunt of it.

It just so happened that we were fishing the same area. When he finally made it up to the grounds, I could see him, and I couldn't believe what I saw. He had all his gear on and it was all iced up. Ice was winged out over the bow and all along the sides of the vessel. He must have had three feet of solid ice on the bow rail. The rule of thumb is that every inch of ice adds about 10,000 pounds to your boat. He must have had over 500,000 pounds of ice on that boat, maybe as much as 750,000 pounds.

And he didn't even know it.

I called him up. "Dad, what are you doing?"

"What are you talking about?" he asked.

"Do you know how much ice you have on that boat right now?"

"No, but the boat feels fine."

When I told him how much ice he had on, he didn't believe me. He couldn't see how much ice he had because the pots were all iced over.

Once he got that top layer of pots off and saw how much ice had built up, it scared him. It scared him big time. Most guys won't let the ice get over six inches, and he had several feet. It looked like a floating ice castle.

My dad had to get that ice off as quickly as he could. They broke out the sledgehammers and jackhammers and went to work. It took him twenty-eight hours to clear the boat.

The incredible thing is that the boat took that much extra weight and was just as steady and responsive as it always was.

It's a hell of a boat. I haven't found anything that I can't put it through, which is kind of scary if you think about it.

I never realized how hard it was to keep money in my pocket until I started making some.

Once I got the *Seabrooke* going in the cod fisheries, I was finally able to pay off all my bills and have a little fun. I bought cars and trucks and snowmobiles. And when the next year's model came out, I bought new ones. Not once or twice, but every single year.

For me it wasn't a status thing. I've never been impressed by people who like to show off their wealth. That wasn't it at all.

I wanted to play.

Because I rarely got a chance to have fun when I was kid, I made up for lost time as an adult. I wanted to go and do all the things I never got to do while I was growing up. There were so many things that I never had the opportunity to experience because I was working all the time. And as a kid, I resented all the things my friends got to do because they didn't have all the responsibilities that I did.

But I didn't have to be resentful anymore. I was like a kid living out his fantasies of what my life would have been like if I hadn't had to work on that damn farm all the time.

I admit, it wasn't the best use of my newfound prosperity, but I couldn't stop myself. I grew up on powdered milk and poached deer. Now I could afford to buy whatever I wanted, whenever I wanted it, and I always wanted more. It was a bit intoxicating.

You'd think with my upbringing I'd be a bit more careful with my money, but I wasn't. I spent freely—especially after I paid off the mortgage on the duplex.

My dad was always in my ear trying to get me to put some money aside, but I didn't listen. We were kicking ass in the cod fishery, and the new program was finally in place. Open-access fishing was finished, a new era was beginning, and the *Seabrooke* was poised to reap the rewards.

Things weren't just looking up. They were looking better than I ever imagined they could. So even though my dad was urging me to be frugal, I wanted what I wanted and couldn't be talked out of it.

What I wanted most was a house. Not just any house, but a mansion. But since they didn't have any for sale in the Walla Walla Valley, I had to build it myself.

I did it right with no expenses spared. The house was forty-two hundred square feet with a master bedroom bigger than our first apartment. Massive kitchen. A huge garage for all my toys. It even had a theater room.

It was a thing of beauty. I wasted no time in filling it will all kinds of crap we didn't really need because I wanted Lisa, Stormee, and Trinidy

With my youngest daughter Trinidy in my custom drag truck with over one thousand horsepower.

to have all the things I didn't have when I was growing up—even if they didn't want them.

Lisa was a simple girl. A country girl. A farmer's daughter. She didn't need a lot of fancy stuff to make her happy. She was a no-frills kind of person. The only thing she wanted was the one thing I couldn't give her—for me to come home and stay home.

After Stormee was born, I'd told Lisa that I'd fish long enough to get us a house and then I'd get out of fishing. Well, it took me more than fifteen years to do it, but I bought myself a house. The problem: It was so big I couldn't afford to stop fishing. My monthly house and car payments were so big that I had no choice but to keep making money.

Lisa didn't want me to fish. That was no secret. She'd always felt that way, and aside from the year I'd spent working for Pepsi, she'd always been the one to compromise.

My drag truck, themed after the *Seabrooke* and crab fishing, with its custom-built aluminum crab steering wheel and door panels with a picture of the *Seabrooke* embroidered into them.

She compromised when I ruptured my appendix. She compromised when the girls were born. She compromised after I lost my finger. She compromised when we bought the *Seabrooke*. But after I missed Stormee's birthday for the tenth year in a row, Lisa decided she wasn't going to put up with it anymore.

Financially it couldn't have come at a worse time. My house payment was five thousand dollars a month. Plus I was paying another fifteen hundred a month in car payments. Lisa didn't understand how I could be making so much money and falling further and further behind. Neither did I.

I wasn't just falling behind, I was going backward. Things got so bad we had to put everything on credit cards. That's when the reality of my situation started to sink in. I was overextended.

I've always been someone who likes to have a good time. Work hard, play hard. That's my credo. Just as I had during my Pepsi days, I started to overdo it with the drinking. It was one thing to celebrate a special occasion or blow off a little steam, but I was drinking every day just to deal with all the pressure of not being able to get out from my mountain of bills. Naturally that led to problems at home.

Lisa and I fought about fishing, we fought about money, we fought about anything and everything, and we fought all the time.

"Why do you even bother coming home?" she asked me. "You're not happy."

She was right. I'd built a big beautiful house for me and my family, but I was hardly ever around to enjoy it. And when I was home, all we did was bicker with each other.

I didn't have an answer for her. Maybe if I did, we could have avoided some heartache. Maybe that's what she wanted me to do: fight to keep our marriage going.

She moved out while I was fishing. When I got wind of what had happened, I called her and begged her to come back.

"Don't do this, Lisa. Don't do it."

But it was too late for words. I'd run out of excuses. There was only one thing that was going to change her mind, and that was for me to quit fishing. But that was the one thing I couldn't do.

Lisa filed the divorce papers for the umpteenth time, and this time she went through with it.

CHAPTER TWELVE

OVER THE SIDE

HERE'S A STORY I'VE NEVER TOLD ANYONE BEFORE.

My drinking buddy on the *Lady Alaska* was Jerry Perkey. One night we went out and got pretty liquored up. When closing time came, we weren't ready to stop. So we headed over to visit Jerry's brother Bob on the boat he was working on, the *Early Dawn*.

It was the end of the season and all the boats were in town. Dutch Harbor isn't all that big, so the boats have to tie up alongside each other. There were so many of them that some docks were ten deep with boats. To get to the boat that was farthest out, you had to cross the other boats. You didn't want your boat to be on the outside spot because you would have to move your boat every time another boat needed to leave. Top producers always got the dockside spots—one of the perks of being a highliner.

The *Early Dawn* was in the last spot out. According to Jerry, Bob had beer on board, so that's where we were headed. To get there we had to cross from boat to boat to boat, which sounds a lot easier than it was. The boats were spaced apart, and the gaps between the boats were anywhere from four to six feet across with nothing but water in between. If you fell, there was nothing to stop your fall.

Drunken fishermen were always falling between the boats. One slip was all it took. The water would be so cold, they usually weren't able to get out. They'd lock up and roll over and that was it.

There were no ladders, nothing to hold on to. Sometimes a fisherman would be able to cling onto a buoy and haul himself out, but if he was too drunk or incoherent to call for help, he died pretty quickly. It's a shitty way to go, but the Bering Sea is the Bering Sea. It doesn't matter if you're in two hundred fathoms of water or twenty. It will get you if you're not careful.

It happened every single season. It was so common that if a guy didn't show up for work in the morning, before the skipper sent someone to go looking for him in town, he'd check to see if there was a floater in the harbor.

The night we went to the *Early Dawn* to drink beer with Bob, we had to cross eight or nine boats. It was dark and icy and the decks were slick. These were boats we'd never been on before, and we didn't know our way around them, especially in the dark.

Jerry and I were pretty tuned up. We weren't really paying attention to what we were doing. We were on a mission, and the only thing on our minds was beer.

I don't know why I looked back. While crossing from one boat to the next, I looked back at Jerry—maybe it was to laugh at something he said or to warn him about something I'd seen—but I lost my balance and fell in the water.

It wasn't one of those deals where everything goes into slow motion and I could see what was happening. No, it happened instantaneously. One second I was cruising along and the next—oh shit—I was under. I went completely underwater, totally submerged.

I hit my head on something as I fell. The deck, a stanchion, I have no idea. My body was all twisted up, so it could have been anything. Whether it was the blow to my head or the shock of plunging into freezing-cold water or the effects of all that alcohol, I became disoriented. Instead of snapping out of my inebriation and taking stock of my situation, I became confused. I couldn't tell which way was up, and I thrashed around in the water like a drowning rat.

A helpless feeling came over me.

I thought I was going to die.

Somehow Jerry was able to reach over and grab me. He pulled me out of the water and dragged me aboard.

I don't know how he did it. It felt like I'd been under for at least a minute, but Jerry said it was only a few seconds.

He hustled me over to the *Early Dawn*. He dried me off and changed me into some clothes that his brother had given him. Then he poured hot coffee down my throat to sober me up and warm my bones.

He saved my life.

At the time, I was more concerned with my dad finding out and getting in trouble than I was with my own safety. But I've thought about that night a lot. It's strange that I've forgotten most of the details, but I remember that feeling of helplessness like it was yesterday. I've never been a quitter, but that's exactly how it felt, like I was giving up. It was as if my brain was taking stock of the fact that at that temperature it was only a matter of minutes before I was dead, so I might as well resign myself to the fact that it's over.

That's it.

End of story.

When I think about that night, I can't get over how quickly it all happened. In the blink of an eye, I was in the water and plummeting toward death. My clock started counting down, and my time left on Earth could be measured in seconds. The thing that gets me every time I let my mind wander down that path is this: If that's how it felt to take a dunk in the harbor, how must it feel to go over the side when you're in the middle of the Bering Sea?

———

"Goddammit, Moose!"

These were the first words out of my mouth. Not "Oh shit!" or "Oh my God!" Not some plea to heaven or cry for help. I didn't panic or freak out. I didn't do any of those things. I cursed when I saw Moose go overboard because I was pissed at him.

As I throttled back the engines and turned the boat around, all I could do was shout obscenities at the man I was going to pluck out of the water.

"Goddammit!"

Down on the deck the crew sprang into action, but I was filled with rage. Going over the side was the most foolish thing a fisherman could do, and the way Moose had done it easily could have been avoided.

All it would have taken was a word of warning, a simple reminder of the fundamentals of fishing, the most basic of safety instructions. But it was too late for that.

A man without a survival suit will last only a few minutes in the Bering Sea. The body wasn't built to withstand immersion in near-freezing water. Not even a mountain of a man like Moose. But somehow the basic facts of cold-weather survival didn't seem to matter. If anyone could survive a bath in the Bering Sea, it was Moose. Of course he would survive.

I never doubted for a second that in a few minutes he'd be standing in the wheelhouse, dripping wet, apologizing for causing such a big ruckus.

"Sorry, Skipper. I don't know what I was thinking . . ."

Then, in a few hours, when we were back at the dock, I'd apologize to him for chewing his ass out so harshly. He'd say he understood, and we'd all have a good laugh about it.

I didn't think we'd lose him. I just didn't believe it was possible. I'd never lost a man before, and I didn't intend to now. I refused to accept it as a possibility. You see it in movies all the time. Some bad shit goes down and all the person can say is, "This can't be happening."

That's how it was with me. Something got disconnected in my brain, went haywire. I was incapable of accepting that Moose was really gone. Even as I sounded the Man Overboard alarm and called the Coast Guard, I couldn't envision a scenario where we wouldn't get him back.

Not because it was a beautiful day or that we'd almost filled our tanks with crab or that we were *this* close to finishing our trip and going back to Dutch Harbor to offload.

Because it was Moose.

That's the kind of man he was. You didn't doubt him. If he said he was going to do something, he did it. He was as dependable a man as I ever met. A tremendous fisherman who came from a great family in Stockton, California. He was the biggest, friendliest, most well-liked guy in all of Dutch Harbor. If he had ever decided to run for mayor of the town, he would have won in a landslide.

Believe me, he was strong. Intensely strong. A guy like that doesn't have to work as hard as others because he can get by on brute strength. But that wasn't Moose. He was a worker. After I challenged him back in the early days of the *Seabrooke,* he became my pacesetter. He led by example and was great with greenhorns. He was one of those guys I knew I could always depend on. That's why I had him on my crew. The first time we'd worked together, brown crabbing out west on the *Lady Alaska,* I thought, *There's a worker. That's the kind of guy I want watching my back.*

When I was putting together the crew of the *Amatuli* for my first stint as captain, Moose had already committed to another boat. But he promised he'd come see me once he finished up. True to his word, he was sitting on the pier with all his gear when they carried me off the *Amatuli,* hand swollen, brain boiling with fever.

"What's happening, Captain?" he hollered as I went by.

That was Moose. Nothing fazed that guy.

Moose had been with me through all the oddball fisheries. He stuck with me while we were trying to figure them out and turn a profit. Even when we were barely making wages, Moose didn't complain. If the boat was working, Moose was working. That's how he was. He loved being out on the sea. It was his home.

By 2009 we finally had a handle on how to fish for bairdi crab. We'd seen the light. It was our third go at it, and we were knocking them dead.

Once we figured out what the hell we were doing, we started pulling not just halfway decent numbers, but some of the best I'd ever seen in that fishery. Normally, getting fifty or sixty crabs per pot was really, really good. We were getting two hundred to the pot. Epic fishing.

It was our first trip of the season, and we were going to fill the boat, which I'd never seen anyone do in that fishery before. You better believe we were stoked.

We'd been getting our ass kicked for almost a week. Rough seas. Miserable wind. If the waves didn't get you, the wind did. The spray lashed you like it held a grudge. I could barely see through the wheelhouse windows what the guys were doing on deck. Everyone was staggering around like a zombie. The conditions weren't bad enough to stop us from working, but it wore the crew down. Mentally we were on high alert, but physically we were on the verge of collapse.

That morning, things changed. The storm broke and we were treated to an absolutely gorgeous sunrise. The seas were flat and calm, a picture-perfect seascape. To say they were "ideal fishing conditions" suggests that we'd seen this kind of scenario before. A mild, sunny day on the Bering Sea is like snow in the Sahara. It just doesn't happen. We didn't get weather like that in the middle of January. Usually January is when the Bering Sea is a cauldron for some of the worst weather in the world. This was something special. A gift from beyond.

The Bering Sea lay before us like a glittering pan of silver, and we just kept pulling up huge pots. We had maybe a day left and then we'd go back to Dutch Harbor with a boat full of crab.

We were on a natural high. Everyone's spirits were soaring. I could see it in the faces of my crew: It doesn't get any better than this. *This* is what you fish for.

Then Moose went over. I still couldn't believe it.

Moose.

Gone.

There are a lot of ways to end up in the water. A rogue wave can slam into the boat and lay it over on its side. That's always bad news if you're topside. A monster wave can wash over the deck and take anything that isn't lashed down with it. Gale force winds can blow a man right off the boat.

It's not always nature. Sometimes your equipment is to blame. A crane might give way or the winches will fail. Turnbuckles break. Harnesses snap. Shit that's supposed to work suddenly doesn't. Winter on the Bering Sea is hell on gear. In spite of all the safety checks, inspections, and evaluations, gear is always breaking down. But nothing broke when Moose went over.

The weather wasn't a factor, either. It was as nice a day as you're ever going to see in those waters.

There's no explanation for why he went over. That's why I was so upset. That's why I was going to chew his ass out.

Just as soon as I got him back on deck.

—◆—

We were setting gear when it happened.

The buoy marks the pot's position. The line that connects the two is called the shot. After the pot is maneuvered into position with the crane, the thrower throws the shot over and the guy on the crane lets the pot go.

When you're the thrower, you do it so many times it becomes second nature. You learn to anticipate the sound of the winches, the movement of the crane, so when the order comes you throw the shot over without thinking about it.

Everything was going smoothly, the way if always went. The order was given, Moose threw the shot, and we dropped the pot.

Part of the shot lay coiled on the deck in a loop. The nautical term for a loop in a line is a *bight,* and like a serpent that's coiled to strike, it can be deadly.

Most of the time when there's line on the deck and the pot's going over, the shot goes over as well. If any part of the shot doesn't go over, you leave it alone and let the pot pull it over. Sometimes it gets hung up on something, but ninety-nine times out of a hundred it gets pulled over by the pot.

From my earliest days as a deckhand I remember learning two rules for deck seamanship: Never step over a line under strain, and never put your foot in a bight. Two simple rules. Words to live by.

Sometimes it's unavoidable. Sometimes people screw up. One season I worked with a deckhand who was always putting the rest of the crew in danger.

When you're setting gear and you go to hook the pot, you have to make sure the block is clear. If it's not clear, the line can pop out of the block. And if the line pops out of the block, the hook can release the line. And if the hook releases the line, the pot prematurely goes over the side. Not only do you lose the pot and everything in it forever, when the pot goes, it takes the line with it. If you get caught in a bight in the line, look out.

Once is an accident. Twice is careless. Three times is a menace.

This kid had done it a couple times, and we were able to trap the line and get the pot back on board. The deck boss told him it better not happen again. One more time and he was done as a deckhand.

He said he was sorry. He said it wouldn't happen again. And then it happened again.

Iced up and setting gear for opilio in the Bering Sea.

The line popped out and the pot went over. As I moved toward the pot, the shot clipped me in the leg and the force of that seven-hundred-pound pot slammed me up against the rail.

One of the other deckhands whipped out his knife and cut the line. We lost the pot, but I was free. The whole thing took about three or four seconds. Another instant and I would have gone over the side.

It jacked me up pretty bad. It's a good thing we were finished with the string because I was done. When I went below and got undressed for bed, I had bruises all over my body.

Normally you hate to see someone lose his job, but I was glad to see that guy get fired. On a fishing boat, we're all in it together. The actions of one impact everyone.

———

Moose had thrown hundreds of thousands of shots during his career as a fisherman. For whatever reason, on that day, on that particular pot, Moose decided to grab what was left of the shot and throw it over. When he lifted his foot up, the line went under strain and the bight grabbed him by the ankle and jerked him over the railing.

It takes longer to describe it than it took to happen. He was there one second, gone the next.

This powerfully strong man who could toy with seven-hundred-pound pots was ripped over the side like a paper doll.

I saw the whole thing happen.

I can still see it.

The crew did exactly what they were supposed to do. We do safety drills once a month. The crew always bitches about them. Give the crew something extra to do, and they act like it's costing them valuable bunk time. But my guys were ready. When the situation went down, everyone did his job.

We had one guy on the crane.

We had another one guy getting in a survival suit.

And we had one guy pointing at the spot where Moose went over, and he never took his eyes off of it.

That's what you do during a Man Overboard drill. Except this wasn't a drill. When we practiced the drill we used a buoy.

I kept waiting for Moose to pop up. I kept combing the water for some sign of him.

Come up, goddammit.

I couldn't change the past. I could only change the future. All I could do was all I could do. That meant getting Moose out of the water, but I couldn't get him out if I didn't know where he was.

I called the Coast Guard and cursed them out. I cursed them for not sending the helicopter fast enough. I cursed the twenty minutes of fly time it would take for them to get to us. I cursed the science that told me that twenty minutes was too late.

Mostly, I cursed Moose.

Moose *knew* better, which is how I knew we were getting him back. It was that simple. Moose knew what he was doing. And so did we. We were ready and able to help. All we needed was an opportunity. We were going to get him back.

The helicopter came on the scene and joined the search. I felt almost euphoric, because I *knew* he was going to pop up any second, and when he did, we'd be there for him.

He's going to pop up . . . He's going to pop up . . . Come on, Moose . . .

I kept my eyes peeled, but the sea played tricks on me. The crazed waves, the cloud shadows, the sameness of the sea. I didn't want to miss him just because I looked in the wrong spot. Maybe a swell took him one direction. Maybe he swam in another. There was no way to know for certain, so I kept searching.

I looked close in. I looked far out. I looked on the other side of the boat in case he was playing a joke and it was all just a stupid stunt that we'd laugh about over beers.

But no.

We looked and we looked.

If I wanted it badly enough, I thought I could will him to appear. I closed my eyes, tried to make myself believe that when I opened them again, he'd be there. And when that didn't work, I cursed again.

Goddammit, Moose!

We looked in all the places we knew we wouldn't find him because he wasn't in any of the places he was supposed to be, and that was right here with us.

We'd keep searching until he was found.

Bob Perkey came up to the wheelhouse with a worried look on his face, one that I knew all too well.

He was worried about something I had yet to grasp myself.

"What now?" I snapped.

"Scott," he said, "it's been eight hours."

"What?"

I looked at the clock. But all it told me was more unbelievable shit.

Eight hours? How was that possible?

In my mind I was thinking it had only been fifteen minutes. I did the math once, twice, three times, trying to make it work. If the math was wrong, than I'd still be right and Moose would pop up . . .

We'd searched for eight hours. Eight hours that went by in a blink of an eye.

That's when it hit me: Moose was gone.

The sea took him from us, and we were never going to see him again.

—◆—

We kept searching. We searched *another* eight hours. At that point we all knew we were searching for a body to give to his family so they could bury him and keep him close. But we couldn't even give them that.

Keith Criner, aka Moose, in front of a full pot of opilio, and sorting king crab with fellow crewmembers on the *Seabrooke* in 2008.

We dissected what had happened. We went over it again and again. But no amount of wondering and what-iffing could change the fact that he was gone and not coming back.

I tried to convince myself that it was his time, because really there was no other explanation. Everyone who makes a living fishing on the Bering Sea knows he may have to pay the ultimate sacrifice. Sometimes we take from the sea, sometimes the sea takes from us.

But I couldn't tell his family that. I couldn't say it to my crew. I couldn't even whisper it to myself when I was all alone in the dark, trying to tally up all my successes and failures, reconcile the good things with the bad. I just couldn't do it.

Moose was gone. Nobody knew why. No one ever would.

It seemed so unfair that with all the risks I'd taken as a man and a captain, all the storms I'd faced, all the demons I'd conquered, that something as ordinary and mundane as a line on the deck would be the thing that got him.

It didn't seem right. It didn't seem fair.

It was something I was going to have to accept. It was something I was going to have to live with for the rest of my life. But I wasn't ready to do that, so I kept looking.

We searched well into the night. Nothing, nothing, nothing. Two thousand square kilometers of nothing.

My guys had been up on deck for more than twenty-four hours. They were physically exhausted and emotionally drained. They were willing to stay up another twenty-four hours if I had asked them to. They'd have stayed up a week for Moose. But it was time to face the brutal reality of life on the Bering Sea. As much as it pained me to admit it, I had to come to terms with the fact that if we hadn't found his body yet, we weren't going to.

It was time to make the call I'd been dreading. I picked up the phone and called the Coast Guard and told them that the *Seabrooke* was going to suspend the search and head back to Dutch Harbor.

It was the hardest thing I've ever had to do in my career. To me, it was the same thing as quitting, and no one can ever tell me different.

The next person I called was my dad. I'd been talking to him throughout the ordeal, asking for his advice, just driving the boat and crying my eyes out because I didn't know what to do.

This time I called him from my stateroom on my cell phone, because I didn't want the rest of the guys to hear what I had to say.

"Come get me," I said. "I need to get off this boat."

CHAPTER THIRTEEN

THE JOURNEY OUT IS THE JOURNEY IN

A NEW DAY BROUGHT RENEWED OPTIMISM AND FRESH HOPE. AT DAWN the following morning, the Coast Guard resumed the search for Moose in their C130 Hercules. The aircraft was designed for long-range search-and-rescue operations. If anyone could find Moose, the Coast Guard could, and they could do it a hell of a lot more efficiently and effectively than we could.

On the way back to town, I tuned in on the radio and listened to the reports. I kept waiting for word that they'd spotted him, but as the updates came in, there was no such news. They couldn't find him.

When the plane started to run low on fuel, they turned around and headed back to base.

Moose was officially lost at sea.

My crewman. My friend.

Gone.

That helpless feeling I had when I fell in the harbor came over me. I remember thinking how much worse it would be in the middle of the Bering Sea. As terrifying as it was to imagine that feeling, being responsible for someone going through it seemed ten thousand times worse.

Aside from a moment of panic when he went over the side and a brief struggle with the line that anchored him to the pot, Moose's suffering was minimal. If I knew Moose, he was probably thinking about us when he went under. In fact I would bet anything that's the case, but that didn't bring me any consolation.

I didn't know what to do. I'd never lost one of my guys before. The smart thing to do would have been to talk to someone who'd been through what I was going through and find a way to move on, but the truth was that I didn't want to move on. As awful as I felt, I needed to hold on to the feeling that I'd let Moose down. He was my responsibility. If I could have switched places with Moose, I would have, and I know the rest of the crew felt the same way. I bargained with God to make it happen. All kinds of crazy thoughts went through my head. I kept waiting to be woken up from this horrible nightmare that wouldn't end.

There had to be something I could do. Some outcome I could influence. Some way I could change this and make it not real.

Then it dawned on me. There was something I could do. It was too late to save Moose, but there was something I could do to make sure this never happened to any of my guys again.

I could quit.

That was it. That was the only solution. I would give up fishing for good.

I made up my mind right then and there. I was officially done with fishing. My dad had told me that he would be on the next flight out of Walla Walla. As soon as he arrived in Dutch Harbor, I'd meet him at the airport and jump on the next plane home.

We headed into port. The sunny skies and balmy weather were already a distant memory. Dark clouds squatted down over Dutch Harbor as we made our approach.

We were all exhausted. The guys looked like robots as they tied the boat up to the dock. I was already thinking, *This is my last trip into Dutch. This is my last time at the processor. This is it.*

On top of the guilt I felt for losing Moose, I felt like I should be say-ing good-bye to everyone, but I didn't want to tip my hand before I told the crew. That wouldn't be right. I'd wait until my dad got here, and when he took over, I'd break the news and say my farewells.

My dad arranged for a grief counselor to meet us on the dock. I had him talk to the guys first. As much as I respected and admired Moose, the deckhands he worked with every day loved him even more. Together we were a family, but they were brothers.

My dad called. I looked at the clock. I figured he was probably in Seattle by then, but there was a chance he was already in Anchorage.

"Where are you?" I asked.

"I'm still at home."

"What? You said . . ."

"I've been on the phone with Moose's family."

Shit. Right. Of course.

"You need to talk to them."

He was right. I made the second-hardest call in my career as a fisher-man and talked to Moose's family.

They were amazing. But I already knew that.

When you spend as much time at sea as we did, you get to know your crew really well. Sometimes too well. You hear about their experiences on other boats. You tell stories about your family and friends back home. You share details you never would have imagined sharing. Sometimes I felt like I knew my crew better than my own family.

So I already knew what great people Moose's family were. What I didn't know was how highly Moose thought of us. Moose's family knew how much fishing meant to him, how much he loved it. He'd told them over and over again that we were his fishing family. When I got on the phone and gave Moose's sister and dad my condolences, they were more worried about *my* well-being than their own. That's the kind of people they are. After talking to them for a few minutes, it was perfectly clear why Moose was so well loved in this community.

And that made it hurt even more. They deserved better than this. They deserved a better captain.

The grief counselor told me not to blame myself. He said it was natural to experience feelings of denial, anger, sadness, and guilt, but that I shouldn't blame myself.

"I'm the captain," I said. "There's no one else to point the finger at."

"It's natural to assign blame," the counselor said, "but there's nothing you could have done."

I heard that again and again and again. Of course there's something I could have done. I made hundreds of decisions during the course of the trip, millions of decisions that added up to Moose going over the side. If just one or two of those decisions had been different, Moose might still be with us today. No one can possibly know in advance that the outcome of those millions of decisions would result in Moose being lost at sea, but to suggest that alternate outcomes weren't possible seemed absurd.

Eventually I figured out that people kept telling me that there was nothing I could have done, not because they believed it, but because it made *them* feel better. But where did that leave me?

I got on the phone to get an ETA from my dad, but he was still in Walla Walla. I couldn't believe it. He told me that he'd spent the afternoon on the phone with Stu, who was still part owner, and dealing with the insurance company, who had a lot of questions. They wanted to know exactly what had happened. He warned me that there would be a gang of paperwork to fill out.

"Fine," I said and slammed the phone on the table.

I felt like my dad was giving me the runaround, and I didn't understand why. Now more than anytime in my life, I needed him, and he kept coming up with excuses.

The phone rang, and I snapped it up, thinking it was my dad, but it wasn't my dad.

It was Lisa.

I didn't know what to say. I didn't need to. Lisa did all the talking. "I'm coming, Scotty. I'm on my way."

———

Every fisherman has a story that he doesn't like to tell.

Mostly he keeps it to himself, but every now and again the urge to tell it will rise up and then it's just a matter of time before it comes spilling out. I never knew my dad had one of those stories—he was the Colonel after all—but a few weeks after we lost Moose, my dad decided to lay it on me.

Back in 1978 my dad was working on a shrimp boat out of Kodiak called the *Mar Del Plata* with his two brothers, Kevin and Dan. Dan had a lease-purchase option on the *Plata*, and they were thinking about partnering up and buying in together. Dan had just bought the *New Venture*, and while he was running that boat, he asked my dad if he wanted to take the *Plata* out shrimp fishing.

My dad had never been in charge of a boat before and was still pretty green, but he jumped at the chance. The Kodiak area had closed, so for his first trip my dad headed down to Chignik Bay, which was about a thirty-six-hour run from Kodiak.

Also on board was his younger brother Kevin and Dan's father-in-law Wayne Lawrence. Wayne wasn't a seasoned fisherman, but he was a gifted mechanic. If anything went wrong with the boat, he would be able to figure it out. They also picked up a guy off the dock who worked in the cannery and had never been fishing before.

They took off and headed down the mainland. They got down to Chignik Bay without incident, but then weather came up. The wind was blowing thirty to thirty-five miles per hour. Kind of rough. Kind of marginal. Not the ideal situation for fishing, especially if you don't know what you're doing.

There were a couple other boats out fishing, so even though he probably should have anchored up and waited for the weather to come down, he set the gear out. My dad felt that if others were fishing, then he ought to be as well.

While the crew was hauling the gear, they got their cables crossed up and were having trouble untangling them and straightening the nets. My dad went to help out, but first he put the boat on autopilot. Back in those days it was called the Iron Mike. The *Mar Del Plata* had a problem with its Iron Mike. In most cases, to engage the autopilot, you simply pulled a lever and locked it in. On the *Mar Del Plata*, every now and again, the Iron Mike would kick out. Between the rough weather and the kinked cables, my dad forgot. He simply set a course, engaged the Iron Mike, and went on deck to help the crew get the nets sorted.

While he was on deck, the Iron Mike kicked out and the boat began to drift. The *Plata* veered off course toward the reef. That's where the shrimp liked to hang out, so they were already fairly close to it. He thought he'd cleared the reef before he went out on deck but the boat was still inside the reef system. And then they hit it.

My dad was on the stern when it happened. He felt a couple of hard bumps underneath him, and then the stern started to rise up. He took off running for the wheelhouse to turn the boat around. He knew they'd hit something, but by the time he got to the wheelhouse it was too late. They were up on the reef and hard aground.

Wayne, the engineer, went down to the engine room and saw water rising. My dad knew he had to get everyone off the boat, but it was easier said than done.

Every time a wave came in, it lifted the boat off the reef and slammed it back down again with enough force to knock everyone off his feet. The boat weighed more than three hundred tons and was wrapped in a skin of quarter-inch plates. Whenever the boat crashed onto the jagged reef, it tore another huge gash in the hull. The boat was like a beer can. It didn't stand a chance against the reef.

My dad put out a Mayday call on the radio and prepared to get everyone off the boat before someone got seriously hurt.

The life raft was sealed in a canister on top of the wheelhouse. To get it my dad had to climb a ladder up the house's backside, open the canister, and pass the life raft down to the others waiting on the deck. The violent crashing of the boat made it a struggle. Every time the boat pitched forward, it smashed down onto the reef with incredible force. The *Mar Del Plata* had transformed from a pokey shrimp boat into a bucking bronco.

My dad got everyone into survival suits, inflated the life raft, and tossed it over the side. With the boat heaving so violently, it was a struggle to get everyone off the boat and into the life raft.

They immediately started to paddle away from the *Plata*. They were very nearly snared in the nets but managed to get away. No matter how hard they tried, they couldn't get away from the sinking vessel. My dad was terrified that the boat's last spasms would swamp the life raft. Finally my dad figured out that their painter was snagged on the *Mar Del Plata*'s railing. As soon as he cut the line, the wind blew them away from the doomed shrimp boat.

There were two other boats in the area that came to my dad's aid. They got everyone safely on board another Kodiak boat called the *Heidi J.* The captain said he'd take them all home—after he finished fishing—and he still had another half-day to go.

So there they were, eight guys jammed into this little boat, and there was nothing my dad could do but sit and wait for the *Heidi J* to finish up. Then it was another thirty-six-hour run back to Kodiak.

Those two days felt like an eternity. Plenty of time to play out in his mind what had happened and what was going to happen. It was a major setback for my dad, both financially and emotionally. It was his first boat, his first opportunity as a captain, and he'd blown it. He'd lost the boat. The *Mar Del Plata* was no more.

My dad couldn't bear the thought of facing his brother. There was a lot of uninsured gear on the boat that his brother was liable for as the leaseholder, and the owner of the boat would certainly sue him. He was ready for an ass chewing of epic proportions. He deserved it. It was his fault. He'd left the wheelhouse unattended. There was no rationalizing away something like that. This was 100 percent on him.

More than anything, he was embarrassed. He'd let his brother down.

My dad wanted to get the hell out of Alaska. He wanted to get on the first plane and never come back.

That first night on the *Heidi J* my dad had his first nightmare about the sinking. He was all alone in a little dinghy out in the middle of the ocean in the pitch dark. He didn't know where he was or what to do. Panic set in, and when he thought he couldn't take anymore, he woke up. He jerked awake in a cold sweat. He had that dream for weeks.

When the *Heidi J* returned to Kodiak, Dan was waiting for my dad on the dock. He had his arms wide open and gave everyone a big hug. He thanked my dad for getting everyone off safely. He told him not to worry about the boat.

"It's just a tool," he said. "It can be replaced. You got everybody home, and that's the important thing."

After a few days in town, my uncle went back out fishing on the *New Venture* and took my dad with him. Neither of them had done much crab fishing before, so they figured it out together. They did a couple of trips and after a couple weeks Dan put my dad in charge.

My uncle knew that if my dad was going to be a successful fisherman he needed to get over his fear of losing the boat. So he basically made my dad get back up on the horse that had thrown him and take the boat out on his own.

My dad was scared shitless. When he pulled out of the harbor, he was covered in sweat. He was terrified of making a mistake that would sink the boat.

By making my dad face his fears, my uncle helped him move past the trauma of losing the boat and become a successful fisherman. My dad fished for another thirty years after the sinking of the *Mar Del Plata*.

Junior and Senior.

But what if my uncle had let him get on a plane? What if my dad had left Alaska?

He never would have come back.

That's why my dad wouldn't let me leave Dutch Harbor. That's why he called Lisa and convinced her to come up and support me.

As much as my dad wanted to be there for me, to help me when I needed him most, he understood that leaving Dutch Harbor would have killed my career.

Even though my mom was furious with him and thought he should have come up to Dutch Harbor to be with me, my dad gave me the same tough love that his brother had given him. He knew that the sooner I got back in the captain's chair, the better off I would be.

He knew what was in my head and in my heart.

He knew because he'd been there.

He knew because he was my dad.

I've never been a particularly patient person. Impulsive to a fault, I'm kind of a fly-by-the-seat-of-my-pants kind of guy. When a storm rolled into Anchorage and grounded all the planes, I nearly lost my shit. I started making plans to go to Anchorage, which would have been foolish at best. My guys were worried about me. They were keeping a close eye on me and stepped in.

They knew I was a mess, so they came up to see me in the wheelhouse. It was kind of like an intervention, an intervention from myself. Bob Perkey spoke for the rest of the crew.

"No matter what happens," Bob began, "we're behind you one hundred percent. It wasn't your fault. There's nothing you could have done or we could have done to prevent this from happening. It was an accident. It happened."

The rest of the guys nodded. Bob continued.

"We want to go back out and finish the season, but we're only going to go if you're running the boat."

As a captain you want your crew to have your back. You trust in each other and you believe in each other and you watch out for each other, but those feelings never get put into words. It's not the fisherman way. For them to stand up for me like that meant a lot.

But it was more than that. They were also standing up *to* me, challenging me. They were saying, "We need you, Captain. Don't let us down."

That was the wake up call I needed.

Being the captain of a boat, any boat, is a great responsibility. I was responsible to the owner. I was responsible to the industry. I was responsible to the crew. If I quit now I wasn't just erasing myself from the equation, I was quitting on *them*.

There was no way I could do that. We'd all been through the same experience together. The crew needed their captain.

It took us a few weeks to collectively pull our shit together, but once we did we were able to get the boat ready in fairly short order and go finish the season.

During that time we received the Coast Guard's verdict: The captain wasn't negligent due to crewman inattention. Instead of giving me comfort, it just pissed me off even more, because the Coast Guard made it sound like it was Moose's fault he went over the side. Why couldn't they put it on me? I was the captain for shit's sake. Isn't the captain ultimately responsible for *everything?*

Once Lisa arrived, I felt a lot better. She calmed me down and helped me get my head on straight. She came with us on our first trip after we lost Moose. When we pulled away from the dock, Lisa stood right by my side. It was the first time she'd ever been out to sea with me, and I was grateful she was there, because I was terrified.

I was shaking. I was sweating. I was so worked up that when we pulled away from the dock, I puked in the trash can.

I settled down a bit, but once we got out to the grounds and started hauling the gear that had been sitting out there, I was a wreck.

I was just so nervous that it would happen again. All my life I was the one who paid the price for my decisions. Whether it was flipping my truck or losing my finger, the consequences were mostly on me. Losing Moose changed that. I was never afraid for myself, and by extension I never worried that much about the crew. All the dumb, reckless things I'd done hadn't seemed so at the time because I wasn't scared. I thought I was in complete control.

I started to see things differently after I lost Moose. It wasn't just me. We were all in this together.

As the boat eased out of the channel and into the sea, Lisa put her arm around my shoulder. She said the guys would be okay. We would be okay. Everything would be okay.

And it was.

CHAPTER FOURTEEN
THE NEXT CHAPTER

WHEN A CRAB FISHERMAN SITS DOWN TO TELL A STORY, IT RARELY HAS a happy ending. There's a reason why so many fishermen's stories are about "the one that got away." I'm doing my best to make sure I'm the exception to the rule.

Shortly after Moose left us, the Discovery Channel contacted me to see if I was interested in appearing on *Deadliest Catch*. I talked it over with my dad, and we decided to give it a shot. The show was already a success when we came aboard, so I can't take any credit for that. As for whether the show is worse off with the *Seabrooke* on it, I'll leave for others to decide.

Deadliest Catch has given me the opportunity to meet a lot of fascinating people on both sides of the camera. Mostly I'm grateful for the way it shows crab fishing for what it is: a difficult and dangerous job performed by some of the hardest-working men in the world. They make me proud to call myself a fisherman.

When I meet people who are fans of the show, they always ask me the same questions. They want to know if the job is as hard as it looks and if the money is really that good.

The answer to both questions is yes.

There's nothing easy about this job. Not one damn thing.

The money is great when it's great, but lots of times it isn't, especially for guys just starting out.

When you're sitting on your couch at home and you see these guys making sixty thousand dollars in two months, it's easy to say, "Hell, I could do that."

The important thing to remember is that not every boat makes that kind of money. If you come up to Alaska with no experience, chances are you're going to end up on a shitty boat. You might only make two grand. You might not make a dime. There are more horror stories than success stories. For the most part, the show focuses on proven moneymakers. Experienced boats with experienced captains and experienced crewmen who have proven themselves. The highliners of the fleet.

There are other highliners who aren't on the show, but there are also a lot of boats that don't make a lot of money. It's no different from any other industry: You're going to have people who are successful and

Signing a T-shirt for a young *Deadliest Catch* fan.

people who just get by and people who lose their shirts. More than likely, if you're a new guy coming into this industry, you're going to end up on a boat that just gets by—or worse. Are you willing to take that risk?

Most people aren't. That's what it's all about, rolling the dice and hoping they come up in your favor.

It's not like the open-access days when every time a boat came in there were twenty guys on the dock looking for work. Some were experienced. Some weren't. Some guys worked on four or five boats in a single season. They'd work on a boat, get fired or quit, party for a while, and when their money ran out they'd jump on another boat. You could do that when there were 375 boats in the fleet.

It's not like that anymore.

People also assume that because of the show, I must be rolling in dough and set for life. That's not the case. One of the first things I did after I joined the show was file for bankruptcy. I finally admitted to myself that being encumbered by that humongous house was a terrible idea. Once I did, I could see how it was killing my marriage. So I got out. When I was finally free and clear of that house, an enormous burden was lifted and I've never looked back.

Lisa, Stormee, Trinidy, and I moved into a much more modest home. There are no theater rooms or palatial suites, but it meets our needs and doesn't exceed our means. We each have a bed to sleep in, and each night there's plenty of food on the table, which occasionally includes cod and crab. As long as there are fish in the sea, the family of a fisherman will never go hungry.

When I'm home, I still cook for Lisa and the girls—if there's room in their busy schedules for Mom and Dad. It's something that gives me a great deal of satisfaction.

I no longer fill my garage with every toy under the sun. I still like to go out and have a good time in the great outdoors, but these days I'm more likely to take a sensible approach and rent my fun.

There are exceptions, though. I still, and always will, have room in the garage for my Harley-Davidson motorcycle. Some things are beyond compromise.

People ask me how long I'll continue fishing. My answer hasn't changed since I started: As long as I can be successful at it, I'll keep doing it. Success comes and goes. I know that better than just about anybody. But things are looking good.

Last year was the best year the *Seabrooke* has ever had. We were profitable in every fishery we participated in, and we had the best cod season in the boat's history. I've had seasons when I caught twice as much total cod, but it took me three times as long to do it. We averaged sixty-five thousand pounds of cod per day, which is exceptional by any standard.

With a little luck we'll continue to have successful seasons with the *Seabrooke*. If so we'll have the boat completely paid off soon and my dad and I will be the sole owners.

That will be a happy day for the Campbell clan.

However, it hasn't all been biscuits and gravy. Last year my dad took a bad fall while he was working in the shop. He had some unexpected complications that were pretty serious, and I left the *Seabrooke* in the middle of the season to be with him. Thankfully he made a full recovery and is back working on his beloved cars.

As a result of being back in Walla Walla during the middle of opilio season, I was home for both of my daughters' birthdays for the first time in, well, ever. Aside from the year I took off to work at Pepsi, I'd never been home for their birthdays. It's hard to write that and not feel like a failure as a father.

If I could do it all over again, would I do anything differently?

You better believe I would.

The first thing I would do differently is make sure I was home when my daughters were born. That was a huge mistake and one of my biggest regrets in life. I should have been there. No ifs, ands, or buts about it.

From left to right—Stormee, Lisa, Trinidy, and me.

Those are two things I'll never be able to experience again, and I believe I'm a lesser man for it. If you have the opportunity to be there when your child is born, do it.

I also wish I had been more honest with Lisa. She never made it a secret that she didn't like me fishing. She didn't care what else I could have done for a living; she just wanted me home and by her side. Instead of sitting down with her and being as honest and forthright with her as she was with me, I lied.

Instead of saying, "Hey, I'm a fisherman, are you going to be with me or not?" I tried to bargain with her. I kept telling her that I was only going to fish one more season.

Or until Stormee reached a certain age.

Or until we achieved a level of financial security that never came.

When I set that dollar amount, I made it seem like it was a reasonable number, an obtainable amount of money that would allow me to walk away from fishing, but I knew that I wouldn't be able to hit that number and would have to fish for at least another year.

I wish I hadn't done that.

I never came clean with Lisa because I was afraid she would leave me. I was afraid if I told her the truth, she would take off. I think that goes back to my issues with my biological mother. But you know what? Lisa isn't my mother. If anyone was like my mother, it was me with my broken promises that I had no intention of keeping in the first place.

In the beginning I should have just said to Lisa, "You can be a fisherman's wife or you can go your own way. What's it going to be?"

Who knows what the outcome would have been if I'd done that. Maybe it would have been better in the long run. Maybe it would have made things easier for her because it would have been her choice.

To be honest, I have no idea how she feels now. I know she doesn't like it that I keep fishing, but she deals with it because it's too late in the game to go back and change who we are. I'm an uneducated man. What else am I gonna do besides fish?

I still enjoy fishing, but I don't know how many more years I have in me. It's still my passion, but I find that I'm less willing to leave home at the beginning of the season and more eager to come home at the end.

I enjoy being home more than I used to, but if I'm home for too long I start to get restless. I don't go salmon tendering anymore, and I don't know how much longer I'll go cod fishing. Cod fishing is a young man's game. I still enjoy coming up to Dutch Harbor and seeing everyone around town. It's hard to believe, but it's been my home away from home for more than twenty years now.

I still love going out on that first trip of the season. I enjoy the competition and challenges of catching cod and crab, but I'll be the first to admit that I don't enjoy it as much as I used to. Once I get the urge to go home, it doesn't go away until we're done.

I think it comes with having achieved everything I set out to do in the industry. I don't have any more aspirations as a captain. I've gone above and beyond what I thought I could achieve in the fisheries. I've pretty much accomplished all of my goals in a relatively short period of time.

I was very aggressive. I was never satisfied. I was never content with a good season, or a good average. As unrealistic as it was, I wanted to surpass myself each and every time. I fished the way my old high school gym teacher wished I would have run races. That was a very valuable lesson. Once the light bulb went on, it never burned out.

It's not like I haven't had my share of fuck ups. I'm the type of guy who has to learn the hard way. I've got to fail before I can succeed—that's how I figure out what I'm trying to accomplish. I get my priorities screwed up. It's probably safe to say they're permanently kinked.

I'm definitely getting mellower with age, especially with my guys. I try to give them a break every now and then. When the season's done, I make sure I say, "Good job."

I'm a lot safer now. If the weather gets too hairy, we'll shut down. I'm more aware of the risks. It's not about proving myself anymore. I've made a name for myself. Everyone knows who I am and what I do. Ask anyone in the fleet about the *Seabrooke* and you'll always hear the same thing: They work fast and they make money. Thanks to *Deadliest Catch*, people who have never set foot in a boat know it, too. There's nothing left for me to prove.

Once you've had those glory seasons, where do you go from there? That's the question I keep asking myself. That's where I'm at now. I'm a top producer, a highliner. Now I'm biding my time until I'm ready to walk away. I don't really know when that will be. I don't have a timeline.

When I lose my passion and no longer enjoy being out on the sea, that's when it will be time to leave. Some people can't walk away from it because they need the money. I hope when my time comes I can afford to walk away. I would hate to die doing something I don't enjoy.

I don't see that happening anytime soon. I know this because when the wind starts blowing and the seas start churning, I get excited.

I can't help it. I love the wild weather. I love the violent seas, the howling winds, the adrenaline pumping through my veins. It's a rush like no other, and I'm not ready to give it up.

I've taken quite a bit of heat over the years for fishing through storms, pushing the envelope, taking the boat out when everyone else is going in. But you know what? I've never lost anyone during a storm. That's when my senses are sharpest and I'm hyper-aware of everything that's going on. My brain and body are on high alert. That's when I'm at my absolute best as a fisherman.

But you know what scares me?

Boredom. Flat seas and calm winds. That's what gets you killed. Because when things get dull, you let your guard down. Your mind drifts. You lose focus. And right when you're thinking happy thoughts about grandma's pies or your girlfriend in the sack, that's when life pulls the rug out from under your feet. I'd rather fish through a storm than in so-called "perfect weather."

The day I catch myself sitting in the wheelhouse bored out of my mind will be my last day as captain of the *Seabrooke*.

I often think about what I'll do when that time comes. I'd like to get into restoring old cars, bringing old death traps back to life. I think I'd enjoy taking something that somebody has thrown away and making it desirable again. I think I would like the challenge of finding the right car, figuring out what it needs, getting it fixed up, and making a little money in the process.

Best of all it's something I could do with my dad. And it would be great to be able work close to home. That's the main thing. I just have to find a way to keep things exciting, and with Lisa, Stormee, and Trinidy in my life, that should never a problem.

When I agreed to do *Deadliest Catch*, I signed a multi-year contract. Prior to that Lisa and I were still taking it one year at a time. Every year we were arguing about whether or not I should keep fishing. The show put that to a stop. In a way *Deadliest Catch* has brought stability to our relationship, because we're not playing that game every year.

The story of my marriage to Lisa is still being written, but in May of 2013 we were officially remarried, and I couldn't be happier about it. We're still trying to figure it out. If it weren't for Lisa, I wouldn't be a fisherman right now.

All I know is that when I needed her most, she was there. In our next chapter together, I hope she can say the same about me.

ACKNOWLEDGMENTS

Scott Campbell, Jr.
I'd like to thank Bill Widing for giving me my first shot running a boat when nobody else would. I would like to thank David Tenzer for putting this project together and Jim Ruland for telling my story the way it needed to be told. Most important, I'd like to thank my family for putting up with me over the years and standing by my side.

Jim Ruland
I'd like to thank the crew of the *Seabrooke* for welcoming me aboard during my visit and the Campbell clan for being such gracious hosts and generous storytellers and taking care of everything I could possibly need, including a ride to the emergency room. Lastly, I'd like to thank my wife Nuvia, *mi sirena y mi amor.*

Index

Seabrooke (boat), *113*, 157,
159, 160
conversion to cod boat, 172–73
photos of, *163*, *165*, *175*, *193*
reputation of, 243
seasonal schedule of, 190–91
in weather, 199, 201–2, *217*
wind meter, 195–96
seasickness, 12
seine net, 14

T
tenders. *See* salmon
tendering
Total Allowable Catch
(TAC), 189

U
Unangan, 98
See also Aleutian Islands

V
Vesser, Wade, 104, 108, 115
on *Arctic Lady*, 101, 102
in Dutch Harbor, 119–20
Vixen (boat), 143, 145

W
Wahl, Fred, 143–44
Walla Walla Valley, 9
Widing, Bill, 132, 133–34, 135
Wilson, Steve, 140–41
wolf eel, 194

ABOUT THE AUTHOR

Captain Scott Campbell, aka Junior, was just twenty-six when he captained his first boat. He's been at the helm of the *Seabrooke* for more than ten years, turning the vessel into one of the top producers in the fleet. Living and working by his motto of "never be satisfied," Junior is always looking for innovative ways to catch crab. He lives in Walla Walla, Washington.

Jim Ruland is a veteran of the Navy and the award-winning author of the short story collection *Big Lonesome*. His essays have appeared in *Granta, Oxford American, Razorcake,* and *Reader's Digest,* and he writes books reviews for the *Los Angeles Times* and *San Diego CityBeat.* He lives in Southern California where he hosts the irreverent reading series Vermin on the Mount.